MULTICONFIGURATIONAL QUANTUM CHEMISTRY

MULTICONFIGURATIONAL QUANTUM CHEMISTRY

By

PROF. BJÖRN O. ROOS

ROLAND LINDH

PER ÅKE MALMQVIST

VALERA VERYAZOV

PER-OLOF WIDMARK

Published by John Wiley & Sons, Inc., Hoboken, New Jersey
Published simultaneously in Canada

For general information on our other products and services or for technical support, please contact our Customer Care Department within the United States at (800) 762-2974, outside the United States at (317) 572-3993 or fax (317) 572-4002.

Wiley also publishes its books in a variety of electronic formats. Some content that appears in print may not be available in electronic formats. For more information about Wiley products, visit our web site at www.wiley.com.

Library of Congress Cataloging-in-Publication Data:

Names: Roos, B. O. (Björn O.) 1937–2010, author. | Lindh, Roland, 1958- author. |
  Malmqvist, Per Åke, 1952- author. | Veryazov, Valera, 1963- author. |
  Widmark, Per-Olof, 1956- author.
Title: Multiconfigurational quantum chemistry / Bjorn Olof Roos, Roland
  Lindh, Per Åke Malmqvist, Valera Veryazov and Per-Olof Widmark.
Description: Hoboken, New Jersey : John Wiley & Sons, 2016. | Includes
  bibliographical references and index.
Identifiers: LCCN 2016010465 (print) | LCCN 2016015079 (ebook) | ISBN
  9780470633465 (cloth) | ISBN 9781119277873 (pdf) | ISBN 9781119277880
  (epub)
Subjects: LCSH: Quantum chemistry–Textbooks.
Classification: LCC QD462 .R66 2016 (print) | LCC QD462 (ebook) | DDC
  541/.28–dc23
LC record available at http://lccn.loc.gov/2016010465

Cover image courtesy of Dr.Valera Veryazov
Set in 10/12pt, TimesLTStd by SPi Global, Chennai, India.

Printed in the United States of America

10 9 8 7 6 5 4

CONTENTS

DEDICATION

In memory of and dedicated to Björn O. Roos, 1937–2010.

The work on this book was started in 2009 by Professor Björn O. Roos. He was in charge of the planning and wrote significant parts before passing away on February 22, 2010. Despite being marked by the deteriorating impact of his condition, Björn spent most of his limited awake time working on this project. Inspired by Björn's enthusiasm and dedication to multiconfigurational wave function theory, we decided to complete the work, as outlined by him, as a testament and a tribute for his contributions in this field.

Thanks Björn!

PREFACE

The intention of this book is to introduce the reader into the multiconfigurational approaches in quantum chemistry. These methods are more difficult to learn to use and there does not exist any textbook in the field that takes the students from the simple Hartree–Fock method to the advanced multireference methods such as multireference configuration interaction (MRCI), or the complete active space self-consistent field (CASSCF) method. The intention is to describe these and other wave function-based methods such that the treatment can be followed by any student with basic knowledge in quantum mechanics and quantum chemistry. Using many illustrative examples, we shall show how these methods can be applied in various areas of chemistry, such as chemical reactions in ground and excited states, transition metal, and other heavy element systems. These methods are based on a well-defined wave function with exact spin and symmetry and are therefore well suited for detailed analysis of various bonding situations. A simple example is the oxygen molecule, which has a $^3\Sigma_g^-$ ground state. Already this label tells us much about the wave function and the electronic structure. It is a triplet state ($S = 1$), it is symmetric around the molecular axis (Σ), it is a *gerade* function, and it is antisymmetric with respect to a mirror plane through the molecular axis. None of these properties are well defined in some methods widely used today. It becomes even worse for the first excited state, $^1\Delta_g$, which cannot be properly described with single configurational methods due to its multiconfigurational character. This failure can have severe consequences in studies of oxygen-containing biological systems. It is true that these wave function-based methods cannot yet be applied to as large systems as can, for example, density functional theory (DFT), but the method development is fast and increases the possibilities for every year.

Computational quantum chemistry is today dominated by the density functional theory and to some extent coupled-cluster-based method. These methods are simple to use and DFT can be applied to larger molecules. They have, however, several drawbacks and failures for crucial areas of applications, such as complex electronic structures, excited states and photochemistry, and heavy element chemistry. Many students learn about the method and how to use it but have often little knowledge about the more advanced wave function-based methods that should preferably be used in such applications.

The intention with this contribution is to demystify the multiconfigurational methods such that students and researchers will understand when and how to use them. Moreover, the multiconfigurational electron structure theory, in association with the molecular orbital picture, has a significant educational and pedagogic value in explaining most chemical processes—the Woodward–Hoffmann rules can only be explained with molecular orbital theory; strict electron density theory will fail.

CONVENTIONS AND UNITS

In this book, we use the conventional systems of units of quantum chemistry, which are the Hartree-based atomic units (au), a set of rational units derived from setting the reduced Planck constant $\hbar = 1$, the electron mass $m_e = 1$, the elementary charge $e = 1$, and the Coulomb constant (4π times the vacuum permittivity) $4\pi\epsilon_0 = 1$, see Tables 1 and 2. The resulting formulae then appear to be dimensionless, and to avoid confusion they are sometimes written in full, that is,

$$-\frac{\hbar^2}{2m_e}\nabla^2 = -\frac{1}{2}\nabla^2 \tag{1}$$

for the kinetic energy term for an electron. Similarly, the electrostatic interaction energy between two electrons can be written with or without the explicit constants:

$$\frac{e^2}{4\pi\epsilon_0 r} = \frac{1}{r}. \tag{2}$$

The first form can be used with any (rational) units. The Bohr, or Bohr radius, and the Hartree, are then used as derived units for length and energy, with symbols a_0 and E_h, respectively. The speed of light is numerically equal to $1/\alpha$, the reciprocal Sommerfeld fine-structure constant, in atomic units.

   Throughout the book, we follow the conventions in Table 3 except where otherwise stated. For example, Ψ will be used as the symbol for a wave function, whereas Φ will be used for configuration state functions. The Hartree–Fock determinant might then in one circumstance be denoted by Ψ if it is the wave function at hand or perhaps as Φ_0 if it is part of an MCSCF expansion.

TABLE 1   One Atomic Unit in Terms of SI Units

Quantity	Symbol	Value
Action	$\hbar$	$1.054\ 571\ 628 \times 10^{-34}$ J s
Mass	m_e	$9.109\ 382\ 15 \times 10^{-31}$ kg
Charge	e	$1.602\ 176\ 487 \times 10^{-19}$ C
Coulomb constant	$4\pi\varepsilon_0$	$4\pi \times 10^7/c^2$ F/m (exact)
Length	a_0	$0.529\ 177\ 208\ 59 \times 10^{-10}$ m
Energy	E_h	$4.359\ 743\ 94 \times 10^{-18}$ J
Electric dipole moment	ea_0	$8.478\ 352\ 81 \times 10^{-30}$ C m
Time	$\hbar/E_h$	$2.418\ 884\ 326\ 505 \times 10^{-17}$ s
Temperature	E_h/k_B	$315\ 774.648$ K

Note: Most of the values above have been taken from web pages of the National Institute of Standards and Technology: http://physics.nist.gov/cuu/.

TABLE 2   Constants and Conversion Factors

Quantity	Symbol	Value
The fine structure constant	α	$0.007\ 297\ 352\ 537\ 6$
Vacuum speed of light	$c = \frac{1}{\alpha}$	$137.035\ 999\ 679$ au
Avogadro's number	N_a	$6.022\ 141\ 79 \times 10^{23}$
Energy	$1E_h =$	$4.359\ 743\ 94 \times 10^{-18}$ J
	$1E_h =$	$627.509\ 469$ kcal/mol
	$1E_h =$	$27.211\ 383\ 86$ eV
	$1E_h =$	$219\ 474.631\ 3705$ cm^{-1}
Electric dipole moment	$1ea_0 =$	$8.478\ 352\ 81 \times 10^{-30}$ C m
	$1ea_0 =$	$0.393\ 430$ Debye

Note: Most of the values above have been taken from web pages of the National Institute of Standards and Technology: http://physics.nist.gov/cuu/.

TABLE 3   Notation Convention Used in This Book

Symbol	Meaning
Ψ	Wave function
Φ	Configuration state function
φ	Orbital (spatial part)
ψ	Spin orbital
σ	Spin wave function
χ	Basis function
ρ	Electronic density
η	Occupation number of (spin)orbital

1

INTRODUCTION

How do we define multiconfigurational (MC) methods? It is simple. In Hartree–Fock (HF) theory and density functional theory (DFT), we describe the wave function with a single Slater determinant. Multiconfigurational wave functions, on the other hand, are constructed as a linear combination of several determinants, or configuration state functions (CSFs)—each CSF is a spin-adapted linear combination of determinants. The MC wave functions also go by the name Configuration Interaction (CI) wave function. A simple example illustrates the situation. The H_2 molecule (centers denoted A and B) equilibrium is well described by a single determinant with a doubly occupied σ orbital:

$$\Psi = (\sigma_g)^2, \tag{1.1}$$

where σ_g is the symmetric combination of the $1s$ atomic hydrogen orbitals ($\sigma_g = \frac{1}{\sqrt{2}}(1s_A + 1s_B)$; the antisymmetric combination is denoted as σ_u). However, if we let the distance between the two atoms increase, the situation becomes more complex. The true wave function for two separated atoms is

$$\Psi \propto (\sigma_g)^2 - (\sigma_u)^2, \tag{1.2}$$

which translates to the electronic structure of the homolytic dissociation products of two radical hydrogens. Two configurations, σ_g and σ_u, are now needed to describe the electronic structure. It is not difficult to understand that at intermediate distances

Multiconfigurational Quantum Chemistry, First Edition.
Björn O. Roos, Roland Lindh, Per Åke Malmqvist, Valera Veryazov, and Per-Olof Widmark.
© 2016 John Wiley & Sons, Inc. Published 2016 by John Wiley & Sons, Inc.

the wave function will vary from Eq. 1.1 to Eq. 1.2, a situation that we can describe with the following wave function:

$$\Psi = C_1(\sigma_g)^2 + C_2(\sigma_u)^2, \tag{1.3}$$

where C_1 and C_2, the so-called CI-coefficients or expansion coefficients, are determined variationally. The two orbitals, σ_g and σ_u, are shown in Figure 1.1, which also gives the occupation numbers (computed as $2(C_1)^2$ and $2(C_2)^2$) at a geometry close to equilibrium. In general, Eq. 1.3 facilitates the description of the electronic structure during any σ bond dissociation, be it homolytic, ionic, or a combination of the two, by adjusting the variational parameters C_1 and C_2 accordingly.

This little example describes the essence of multiconfigurational quantum chemistry. By introducing several CSFs in the expansion of the wave function, we can describe the electronic structure for a more general situation than those where the wave function is dominated by a single determinant. Optimizing the orbitals and the expansion coefficients, simultaneously, defines the approach and results in a wave function that is qualitatively correct for the problem we are studying (e.g., the dissociation of a chemical bond as the example above illustrates). It remains to describe the effect of dynamic electron correlation, which is not more included in this approach than it is in the HF method.

The MC approach is almost as old as quantum chemistry itself. Maybe one could consider the Heitler–London wave function [1] as the first multiconfigurational wave function because it can be written in the form given by Eq. 1.2. However, the first multiconfigurational (MC) SCF calculation was probably performed by Hartree and coworkers [2]. They realized that for the ^1S state of the oxygen atom, there where two possible configurations, $2s^2 2p^4$ and $2p^6$, and constructed the two configurational wave function:

$$\Psi = C_1 \Phi(2s^2 2p^4) + C_2 \Phi(2p^6). \tag{1.4}$$

The atomic orbitals were determined (numerically) together with the two expansion coefficients. Similar MCSCF calculations on atoms and negative ions were simultaneously performed in Kaunas, Lithuania, by Jucys [3]. The possibility was actually

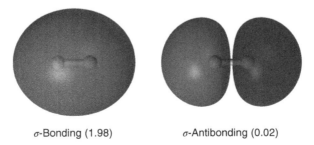

σ-Bonding (1.98) σ-Antibonding (0.02)

Figure 1.1 The σ and σ^* orbitals and associated occupation numbers in the H_2 molecule at the equilibrium geometry.

suggested already in 1934 in the book by Frenkel [4]. Further progress was only possible with the advent of the computer. Wahl and Das developed the *Optimized Valence Configuration (OVC) Approach*, which was applied to diatomic and some triatomic molecules [5, 6].

An important methodological step forward was the formulation of the *Extended Brillouin's (Brillouin, Levy, Berthier) theorem* by Levy and Berthier [7]. This theorem states that for any CI wave function, which is stationary with respect to orbital rotations, we have

$$\langle \Psi | \hat{H} \hat{E}_{ai}^{-} | \Psi \rangle = 0, \tag{1.5}$$

where $\hat{E}_{ai}^{-}$ is an operator (see Eq. 9.32) that gives a wave function $\hat{E}_{ai}^{-} | \Psi \rangle$ where the orbitals φ_i and φ_a have been interchanged by a rotation. The theorem is an extension to the multiconfigurational regime of the Brillouin theorem, which gives the corresponding condition for an optimized HF wave function. A forerunner to the BLB theorem can actually be found already in Löwdin's 1955 article [8, 9].

The early MCSCF calculations were tedious and often difficult to converge. The methods used were based on an extension of the HF theory formulated for open shells by Roothaan [10]. An important paradigm change came with the *Super-CI* method, which was directly based on the BLB theorem [11]. One of the first modern formulations of the MCSCF optimization problem was given by Hinze [12]. He also introduced what may be called an approximate second-order (Newton–Raphson) procedure based on the partitioning: $U = 1 + T$, where U is the unitary transformation matrix for the orbitals and T is an anti-Hermitian matrix. This was later to become $U = exp(T)$. The full exponential formulation of the orbital and CI optimization problem was given by Dalgaard and Jørgensen [13]. Variations in orbitals and CI coefficients were described through unitary rotations expressed as the exponential of anti-Hermitian matrices. They formulated a full second-order optimization procedure (Newton–Raphson, NR), which has since then become the standard. Other methods (e.g., the Super-CI method) can be considered as approximations to the NR approach.

One of the problems that the early applications of the MCSCF method faced was the construction of the wave function. It was necessary to keep it short in order to make the calculations feasible. Thus, one had to decide beforehand which where the most important CSFs to include in the CI expansion. Even if this is quite simple in a molecule like H_2, it quickly becomes ambiguous for larger systems. However, the development of more efficient techniques to solve large CI problems made another approach possible. Instead of having to choose individual CSFs, one could choose only the orbitals that were involved and then make a full CI expansion in this (small) orbital space. In 1976, Ruedenberg introduced the *orbital reaction space* in which a complete CI expansion was used (in principle). All orbitals were optimized—the *Fully Optimized Reaction Space—FORS* [14].

An important prerequisite for such an approach was the possibility to solve large CI expansions. A first step was taken with the introduction of the *Direct CI* method in 1972 [15]. This method solved the problem of performing large-scale SDCI calculations with a closed-shell reference wave function. It was not useful for MCSCF, where a more general approach is needed that allows an arbitrary number of open shells and all possible spin-couplings. The generalization of the direct CI method to such cases was made by Paldus and Shavitt through the *Graphical Unitary Group Approach (GUGA)*. Two papers by Shavitt explained how to compute CI coupling coefficients using GUGA [16, 17]. Shavitt's approach was directly applicable to full CI calculations. It formed the basis for the development of the *Complete Active Space (CAS) SCF* method, which has become the standard for performing MCSCF calculations [18, 19].

However, an MCSCF calculation only solves part of the problem—it can formulate a qualitatively correct wave function by the inclusion of the so-called static electron correlation. This determines the larger part of the wave function. For a quantitative correct picture, we need also to include dynamic electron correlation and its contribution to the total electronic energy. We devote a substantial part of the book to describe different methods that can be used. In particular, we concentrate on second-order perturbation theory with a CASSCF reference function (CASPT2). This method has proven to be accurate in many applications also for large molecules where other methods, such as MRCI or coupled cluster, cannot be used. The combination CASSCF/CASPT2 is the main computational tool to be discussed and illustrated in several applications.

This book mainly discusses the multiconfigurational approach in quantum chemistry; it includes discussions about the modern computational methods such as Hartree–Fock theory, perturbation theory, and various configuration interaction methods. Here, the main emphasis is not on technical details but the aim is to describe the methods, such that critical comparisons between the various approaches can be made. It also includes sections about the mathematical tools that are used and many different types of applications. For the applications presented in the last chapter of this book, the emphasis is on the practical problems associated with using the CASSCF/CASPT2 methods. It is hoped that the reader after finishing the book will have arrived at a deeper understanding of the CASSCF/CASPT2 approaches and will be able to use them with a critical mind.

1.1 REFERENCES

[1] Heitler W, London F. Wechselwirkung neutraler Atome und homopolare Bindung nach der Quantenmechanik. Z Phys 1927;44:455–472.

[2] Hartree DR, Hartree W, Swirles B. Self-consistent field, including exchange and superposition of configurations, with some results for oxygen. Philos Trans R Soc London, Ser A 1939;238:229–247.

[3] Jucys A. Self-consistent field with exchange for carbon. Proc R Soc London, Ser A 1939;173:59–67.

[4] Frenkel J. Wave Mechanics, Advanced General Theory. Oxford: Clarendon Press; 1934.

[5] Das G, Wahl AC. Extended Hartree-Fock wavefunctions: optimized valence configurations for H_2 and Li_2, optimized double configurations for F_2. J Chem Phys 1966; 44:87–96.

[6] Wahl AC, Das G. The multiconfiguration self-consistent field method. In: Schaefer HF III, editor. Methods of Electronic Structure Theory. New York: Plenum Press; 1977. p. 51.

[7] Levy B, Berthier G. Generalized Brillouin theorem for multiconfigurational SCF theories. Int J Quantum Chem 1968;2:307–319.

[8] Löwdin PO. Quantum theory of many-particle systems. I. Physical interpretations by means of density matrices, natural spin-orbitals, and convergence problems in the method of configurational interaction. Phys Rev 1955;97:1474–1489.

[9] Roos BO. Perspective on "Quantum theory of many-particle systems I, II, and III" by Löwdin PO, [Phys Rev 1995;97:1474–1520]. Theor Chem Acc 2000;103:228–230.

[10] Roothaan CCJ. Self-consistent field theory for open shells of electronic systems. Rev Mod Phys 1960;32:179–185.

[11] Grein F, Chang TC. Multiconfiguration wavefunctions obtained by application of the generalized Brillouin theorem. Chem Phys Lett 1971;12:44–48.

[12] Hinze J. MC-SCF. I. The multi-configuration self-consistent-field method. J Chem Phys 1973;59:6424–6432.

[13] Dalgaard E, Jørgensen P. Optimization of orbitals for multiconfigurational reference states. J Chem Phys 1978;69:3833–3844.

[14] Ruedenberg K, Sundberg KR. MCSCF studies of chemical reactions. I. Natural reaction orbitals and localized reaction orbitals. In: eds Calais JL, Goscinski O, Linderberg J, Öhrn Y, editors. Quantum Science; Methods and Structure. New York: Plenum Press; 1976. p. 505.

[15] Roos BO. A new method for large-scale CI calculations. Chem Phys Lett 1972;15:153–159.

[16] Shavitt I. Graph theoretical concepts for the unitary group approach to the many-electron correlation problem. Int J Quantum Chem 1977;12:131–148.

[17] Shavitt I. Matrix element evaluation in the unitary group approach to the electron correlation problem. Int J Quantum Chem 1978;14:5–32.

[18] Roos BO, Taylor PR, Siegbahn PEM. A complete active space SCF method (CASSCF) using a density matrix formulated super-CI approach. Chem Phys 1980;48:157–173.

[19] Roos BO. The complete active space self-consistent field method and its applications in electronic structure calculations. In: Lawley KP, editor. Advances in Chemical Physics; Ab Initio Methods in Quantum Chemistry - II. Chichester: John Wiley & Sons, Ltd; 1987. p. 399.

2

MATHEMATICAL BACKGROUND

2.1 INTRODUCTION

From a basic point of view, orbitals are not *the* orbitals of some electron system, but they are *a* convenient set of one-electron basis functions. They may, or may not, solve some differential equations.

The ones that are most used in contemporary Quantum Chemistry are described in more technical detail further on. Here we just mention a few basic properties, and some mathematical facts and notations that come in handy. Later in the book it also describes the methods whereby the wave functions, which are detailed descriptions of the quantum states, can be approximated.

This chapter is also concerned with the practical methods to represent the many-electron wave functions and operators that enter the equations of quantum chemistry, specifically for bound molecular states.

It ends with some of the tools used to get properties and statistics out from multi-configurational wave functions. They all turn out to be, essentially, "matrix elements," computed from linear combinations of a basic kind of such matrix elements: the density matrices.

2.2 CONVENIENT MATRIX ALGEBRA

There are numerous cases where linear or multilinear relations are used. Formulas may be written and handled in a very compact form, as in the case of orbitals being

Multiconfigurational Quantum Chemistry, First Edition.
Björn O. Roos, Roland Lindh, Per Åke Malmqvist, Valera Veryazov, and Per-Olof Widmark.
© 2016 John Wiley & Sons, Inc. Published 2016 by John Wiley & Sons, Inc.

built from simpler basis functions (or other orbitals). The one-particle basis functions $\chi_k, k = 1, \ldots, n$ are arranged in a row vector, formally a $1 \times n$ matrix, and the coefficients for their linear combinations, often called MO coefficients, in an $n \times m$ matrix $\mathbf{C}$, so that the orbitals $\{\phi_k\}_{k=1}^m$, can be written very concisely as a matrix product:

$$(\phi_1, \phi_2, \ldots, \phi_m) = (\chi_1, \chi_2, \ldots, \chi_n) \begin{pmatrix} C_{11} & C_{12} & \cdots & C_{1m} \\ C_{21} & C_{22} & \cdots & C_{2m} \\ \cdots & \cdots & \cdots & \cdots \\ C_{n1} & C_{n2} & \cdots & C_{nm} \end{pmatrix} \tag{2.1}$$

that is, $\phi = \chi\mathbf{C}$.

As an example, the orbital optimization procedure in a Quantum Chemistry program is frequently carried out by matrix operations such as, for example, the matrix exponential function, as shown in Chapter 9. The same approach can be extended also to handle many-particle wave functions.

Mathematically, the Schrödinger equation is usually studied as a partial differential equation, while computational work is done using basis function expansions in one form or another. The model assumption is that the wave functions lie in a Hilbert space, which contains the square-integrable functions, and also the limits of any convergent sequence of such functions: molecular orbitals, for example, would have to be normalizable, with a norm that is related to the usual scalar product:

$$\langle \phi_k | \phi_l \rangle \overset{\text{def}}{=} \int_{\mathcal{R}^3} \phi_k^*(\mathbf{r}) \phi_l(\mathbf{r}) \, d\mathbf{r}^3 \quad \text{and} \quad \|\phi_k\| \overset{\text{def}}{=} \sqrt{\langle \phi_k | \phi_k \rangle}.$$

This space of orbitals, together with the norm and scalar product, is called $L^2(\mathcal{R}^3)$, a separable Hilbert space, which means that it can be represented by an infinite orthonormal basis set. Such a basis should be ordered, and calculations carried out using the first N basis functions would be arbitrarily good approximations to the exact result if N is large enough. There are some extra considerations, dependent on the purpose: for solving differential equations, not only the wave functions but also their derivatives must be representable in the basis, and so a smaller Hilbert space can be used. For quantum chemistry, this can be regarded as requiring that the expectation value of the kinetic energy operator should be finite, and the wave function should then lie in a subspace of $L^2(\mathcal{R}^3)$, where also $\|\nabla\phi\|$ is finite (a so-called Sobolev space). While this is naturally fulfilled for most kinds of bases, it is not always so, for example, for finite element functions, wavelets, and in complete generality, issues such as completeness, convergence rate, and accuracy can be complicated.

Operators tend to be positions, partial derivatives, or functions of these. State vectors are usually wave functions with position variables, with spin represented by additional indices such as α or β. Examples are as follows:

$$\hat{\nabla} = \mathbf{e}_x \frac{\partial}{\partial x} + \mathbf{e}_y \frac{\partial}{\partial y} + \mathbf{e}_z \frac{\partial}{\partial z},$$

$$\hat{\nabla} \cdot \mathbf{r} = \text{div}(\mathbf{r}) + x\frac{\partial}{\partial x} + \cdots + z\frac{\partial}{\partial z} = 3 + \mathbf{r} \cdot \hat{\nabla},$$

$$\hat{\mathbf{s}} = \mathbf{e}_x \hat{s}_x + \mathbf{e}_y \hat{s}_y + \mathbf{e}_z \hat{s}_z,$$

$$\hat{s}_z \phi_\alpha = \frac{1}{2}\phi_\alpha \quad \hat{s}_z \phi_\beta = -\frac{1}{2}\phi_\beta.$$

We note that there are vector operators, which act by producing a vector with elements that are wave functions: $\hat{\nabla} f(\mathbf{r}) = (\partial f/\partial x, \partial f/\partial y, \partial f/\partial z)$ in the natural way. We also note that operators are defined by their effect when acting on a function. One thing to look up for is using, for example, polar coordinates, a partial derivative or a $\hat{\nabla}$ operator may act on a vector expressed with the basis vectors $\mathbf{e}_r, \mathbf{e}_\theta, \mathbf{e}_\phi$. These are not constant vectors, and their derivatives yield extra terms, for example, for angular momentum operators. We also note that order matters—operators are usually not commutative, as seen for $\hat{\nabla}$ and $\mathbf{r}$. For any two operators $\hat{A}$ and $\hat{B}$, one defines the commutator $[\hat{A}, \hat{B}]$ and the anticommutator $[\hat{A}, \hat{B}]_+$:

$$[\hat{A}, \hat{B}] = \hat{A}\hat{B} - \hat{B}\hat{A} \quad \text{and} \quad [\hat{A}, \hat{B}]_+ = \hat{A}\hat{B} + \hat{B}\hat{A}. \tag{2.2}$$

The so-called Dirac notation, or bra-ket notation, is common and very useful. It is simply explained by starting with a vector space scalar product, which can be, for example,

$$\langle v | \hat{D} \, u \rangle,$$

where u and v are some vectors, and $\hat{D}$ is some linear operator in that space. This can be an infinite-dimensional Hilbert space, like the Sobolev spaces, but this notation can be used for any general vector space. Dirac notation implies that another vertical bar symbol is introduced, and the syntax is then that this is a triple of the following constituents:

- A vector, written $|u\rangle$, called a "ket vector"
- An operator, as before written as $\hat{D}$
- A linear functional, written $\langle v|$, with the property that when "acting" on a ket vector, it produces a scalar value, usually complex.

The linear functional is an element of a linear vector space, formally the "dual space" of the ket space. It is called a "bra" vector or a "bra functional." For a Hilbert space, its dual is also a Hilbert space, isomorphic with the ket space, and for the usual function spaces, they can be simply identified without causing any problems. The actual functions, used in integrals, can be used both as ket and bra vectors just by complex conjugation.

This is not entirely true for all spaces, or when "Dirac δ distributions" are used in the integrals. However, we usually feel free to use Dirac distributions as if they were functions, usually arising from the "resolution of the identity," which starts with the well-known formula

$$|u\rangle = \sum_{k=1}^{N} |e_k\rangle\langle e_k|u\rangle,$$

which is true for any finite vector space with an orthonormal basis $\{e_k\}_{k=1}^{N}$. It is also true for any so-called "separable" infinite Hilbert space, which is simply those in which there are infinite orthonormal bases $\{e_k\}_{k=1}^{\infty}$. This is essentially all spaces that we have reason to use in Quantum Chemistry! There is just a couple of caveats: one must remember that the scalar product, and the norm, are then written in terms of integrals, which do not distinguish between any functions that differ only in isolated points. Function values in isolated points are not "useable," and Dirac δ distributions do not formally have any place in the formalism. However, this particular problem disappears with the simple stratagem of regarding expressions involving Dirac distributions as constructs that imply the use of a "mollifier." In this context it is just a parametrized function, which has the property of being nonnegative, bounded, zero for $|x| > |\epsilon|$, and having the integral 1 if integrated from any negative to any positive value, and with the evaluation rule that the limit $\epsilon \to 0^+$ is to be taken finally. This allows us to define a unit operator as

$$\sum_{k=1}^{\infty} |e_k\rangle\langle e_k| = \hat{1} \tag{2.3}$$

and translate it to functions as

$$\sum_{k=1}^{\infty} \phi_k(x)\phi_k^*(x') = \delta(x - x'). \tag{2.4}$$

The multivariate extensions are obvious.

This also allows us to represent operators, by writing

$$\hat{A} = \hat{1}\,\hat{A}\,\hat{1} = \sum_{k=1}^{\infty} |e_k\rangle\langle e_k|\hat{A} \sum_{l=1}^{\infty} |e_l\rangle\langle e_l| = \sum_{kl} |e_k\rangle A_{kl}\langle e_l|, \tag{2.5}$$

where $A_{kl} = \langle e_k|\hat{A}|e_l\rangle$ is a matrix representation of the operator, and we also get a representation in terms of basis functions as a so-called integral kernel,

$$A(x, x') = \sum_{kl} \phi_k(x)A_{kl}\phi_l^*(x'),$$

which is to be used as

$$\hat{A}|\psi\rangle = \int_{-\infty}^{\infty} A(x, x')\psi(x')\,dx'. \tag{2.6}$$

It is thus seen that, with an innocent abuse of notation, we can alternately implement the "bra-ket" notation in terms of matrices and sums (although in general infinite ones), or integral kernels (with some suitable handling of any differential operators). This brings us to another advantage: the notation is essentially the same if some of the scalar products are sums over distinct values, rather than integrals. It is, for example, no problem to use the electron spin together with the position variable, although the spin is binary (α or β) while the position is, for example, a triple of Cartesian coordinates.

We are mostly used to the "standard" Gaussian basis sets. For these, it may be pointed out that they are not members of a complete basis set for $L^2(\mathcal{R}^3)$. Such basis sets do exist, for example, the complete set of harmonic oscillator eigenfunctions. That is a fixed sequence of basis functions, and for $N < M$, the span of the N first functions is a proper subspace of the span of the M first functions. Instead, for the Gaussian bases, one can devise sequences of different basis sets (larger and larger but not obtained by merely adding functions), such that each basis set allows construction of an approximate wave function, and the sequence of approximate wave functions converges (also pointwise) to a given wave function. Formal requirements for such a sequence to be complete have been described by Feller and Ruedenberg [1].

For the representation of orbitals, that is, single-particle functions, practical and theoretical aspects on the choice and use of basis sets are dealt with in Chapter 6. Here, we now leave those considerations behind, and merely assume that in any specific calculation, there is a "large enough" one-electron basis that is used in forming a large but finite set of orthogonal basis functions, the molecular orbitals (abbreviated MOs), and that these can be used as an approximation to the complete basis.

2.3 MANY-ELECTRON BASIS FUNCTIONS

We also need a set of basis functions for the many-electron wave functions. In the usual wave function representation, all terms in the Hamiltonian, as well as any additional operators that represent perturbations and/or properties that should be computed, are one- or two-particle functions. We must be able to represent these faithfully. For the moment, we assume this to be true within some acceptable accuracy, even for a finite basis. For a many-electron basis that contains all products of one-electron basis function, the only problem is to handle the two-electron terms of the Hamiltonian. It turns out that, for example, the Coulomb interaction, in spite of going to infinity when particles coalesce, is also representable in such a basis. Special considerations are needed, for example, for some terms used in relativistic Quantum Chemistry.

For electrons, as for any indistinguishable fermions, it is known that the wave function is antisymmetric: it will change sign if any two electron variables are interchanged. A typical such function is the *Slater determinant* (SD), which for any set of

N one-electron functions forms an antisymmetric product in N variables, for example, with $N = 3$:

$$|ijk\rangle \stackrel{\text{def}}{=} \frac{1}{\sqrt{3!}} \begin{vmatrix} \psi_i(x_1) & \psi_j(x_1) & \psi_k(x_1) \\ \psi_i(x_2) & \psi_j(x_2) & \psi_k(x_2) \\ \psi_i(x_3) & \psi_j(x_3) & \psi_k(x_3) \end{vmatrix}. \tag{2.7}$$

The determinant functions can also be called an antisymmetric tensor products of orbitals.

Moreover, the electron spins play an important part, and must be somehow represented. In relativistic quantum mechanics, orbitals are two- or four-component quantities, called "spinors." Also nonrelativistically, the two-component form is a good way of treating spin, especially for magnetic interactions. A two-component spin-orbital or spinor basis function can be written as

$$\psi_k(\mathbf{r}) = \begin{pmatrix} \phi_{k1}(\mathbf{r}) \\ \phi_{k2}(\mathbf{r}) \end{pmatrix}, \tag{2.8}$$

where the two components indicate the complex amplitude of an α spin and β spin, respectively. This is not quite suitable for writing products: We would need to form four different components for the product of two spinors, and eight components for three, etc. One also wants to be able to deal with spatial and spin separately, and then the more convenient way is to use two ordinary one-electron bases of real or complex functions. The one-electron bases considered in this chapter are orbital functions; that these may be, in turn, linear combinations of some common (typically Gaussian) basis is immaterial. The orbitals could also in some applications be, for example, numerical tables of values on a grid. The basis set for α and for β spin are often the same, and in this case they are conveniently treated as if they were the product of a spatial part and a spin function, which in that case is shown as $\alpha(1)$, as a function of particle 1, etc.

Consider a wave function that is written as the antisymmetrized product of spin-orbitals. The spin part is formally written as a function $\alpha(j)$ of $\beta(j)$, where $j = 1, 2, \cdots$ is a label enumerating the particles. Similarly, the spatial orbital part is written $\phi_i(j)$, meaning that function nr. i is used to describe particle nr. j. We already know that such a wave function is called a Slater Determinant (SD), for example,

$$\frac{1}{\sqrt{2}} \begin{vmatrix} \phi_1(1)\alpha(1) & \phi_2(1)\beta(1) \\ \phi_1(2)\alpha(2) & \phi_2(2)\beta(2) \end{vmatrix} = \frac{1}{\sqrt{2}}(\phi_1(1)\alpha(1)\phi_2(2)\beta(2) - \phi_1(2)\alpha(2)\phi_2(1)\beta(1))$$

for two electrons.

We write such an SD in the abbreviated form $|\phi_1\bar{\phi}_2\rangle$ or even as short as just $|1\bar{2}\rangle$. We note that the normalizer $1/\sqrt{2}$ is implied in the short form, that in the shortest form the numbers indicate the orbital labels, and that β spin is indicated by an overbar.

Knowing the rules for evaluating a determinant, we note that the function is indeed antisymmetric if we interchange particle indices, and also if we interchange the functions (remembering to interchange both spatial and spin function!).

We also note that (in this case) there is no particular symmetry if only the spatial or only the spin functions are interchanged. If one prefers wave functions constructed such that they have specific spatial symmetry *and* spin symmetry, this is perfectly possible: the determinant above must then be written as the sum of two functions, one is antisymmetric in space and symmetric in spin, and would be called a "triplet" function; the other is vice versa symmetric in space, antisymmetric in spin, and is a "singlet."

Given any N-electron wave function Ψ, there exists an orthonormal basis $\{\phi_k\}_{k=1}^{\infty}, k = 1, 2, \ldots$ of *natural spin-orbitals*, with very special qualities. Each is associated with a *natural spin-orbital occupation*, $0 \leq n_k \leq 1$. The sum of the occupation numbers is N.

Consider an approximation to Ψ by Ψ', a wave function formed as a Full CI using any finite, orthonormal set of spin-orbitals $\{\phi_k'\}_{k=1}^{M}$. Let this function be defined as the projection of Ψ on the Full CI space spanned by all Slater determinant terms that can be formed using the orbitals $\{\phi_k'\}_{k=1}$.

A theorem by Davidson [2] states that choosing $\{\phi_k'\}_{k=1}^{M}$ to consist of those M natural orbitals that have the largest natural occupations is optimal, in the sense that it gives the largest possible overlap with the exact wave function.

- For any given finite set of wave functions, it is possible to construct the *best algebraic approximation* to the linear span of this set. Assume that we wish to allow at most N (e.g., 100) basis functions, and that the quality number (defining "best") is that of maximum overlap. We have already seen a solution, above. For many similar problems, there are similar well-known methods for constructing a sequence of basis functions, associated with a decreasing sequence of positive quality numbers, such that the problem is uniquely solved by picking the first N functions. This is usually some variant of the singular value decomposition (SVD).

- The SVD takes a data set in the form of a large $m \times n$ matrix A_{kl}, which requires on the face of it mn data values. It constructs a decomposition

$$A_{kl} = \sum_I \lambda_I u_k^I v_l^I,$$

where the "singular values" $\{\lambda_I\}_{I=1}^{N}$ are nonnegative, nonincreasing, and $N \leq \max(m, n)$. In fact, the hope is that the singular values are *rapidly* decreasing, and then in practice N can be chosen to be much less than the limit, $\max(m, n)$, and then the error in this representation can be directly related to the size of the neglected singular values. Try $m = n = 10^6$ and $N = 1000$, for instance. In the particular case that A is symmetric (Hermitian) with nonnegative eigenvalues, the usual orthonormal diagonalization will provide an SVD, but the general decomposition exists for any matrix.

- An example is given in Chapter 6, where this is used to construct the so-called ANO basis sets. But this principle is used in all heavy applications of numerical linear algebra, when possible, including in Quantum Chemistry, of course.

- For bound states, there is the well-known variational theorem, which states that by solving the Schrödinger equation projected on a fixed basis, and ordering the eigenvalues (i.e., energies) from below and increasing, the approximated energies are upper bounds to the true energies. It is also possible to compute lower bounds, and estimates of the overlap between the approximated and the true eigenfunctions. This is the solid ground for computing wave functions of quantum systems by variation. In its oldest form, it is called the "Rayleigh–Ritz" variation theorem.

If the matrix representation of the Hamiltonian consists of integrals over the wave function basis, then the eigenvalues of the Hamiltonian matrix are accurate upper bounds to the eigenvalues of the Schrödinger equation. The variation theorem is then a useful tool for devising basis sets. If the Schrödinger equation is rewritten in other, formally equivalent forms, this variational property may be lost. Nevertheless, approaches such as truncated Coupled Cluster methods can at least be cast in variational forms that do not give bounds for the Schrödinger eigenvalues but which are nevertheless very useful, for example, for formulating consistent forces and other derivatives. Also, approximating, for example, products or inverses of operators by the corresponding finite matrix algebra often yields controllable approximations.

In practice, the basis sets are usually too small anyway to allow the *absolute* or *total* energies to be used in the variational theorem to deduce the accuracy of the wave function. Good properties require basis sets that have been optimized for that purpose.

In the following, the quality of the basis set is disregarded; it is assumed to be "good enough" for its purpose. The usefulness for wave functions follows from the well-known fact that a complete basis for an n-particle wave function can be constructed as the "tensor" or "outer" product of the one-particle basis functions. In the case of fermions like electrons, this wave function basis can be "adapted" to the various symmetries that the particles obey, primarily that they are antisymmetric under particle interchange, and if spin-dependent interactions can be ignored, that the total spin is a good quantum number. The antisymmetry gives naturally a basis of Slater Determinants.

2.4 PROBABILITY BASICS

The following conventions are generally used in probability theory: The probability is expressed as a function, often written as $P(A)$, where A is an event, which can be written in words as a statement, such as the probability that "particle 1 and particle 2 are closer than 1 Å," usually expressed using variables, that is, $P(|\mathbf{r}_1 - \mathbf{r}_2| \leq s)$. Most probably, one writes "Let $P(s) = P(|\mathbf{r}_1 - \mathbf{r}_2| \leq s)$," so in the same context, one can use just $P(s)$ as a function. In this example, s is an ordinary real variable, and it is rarely restricted, that is, it may be negative, and it can be arbitrarily large. In this case, if $s < 0$, it just means that the statement is never true, and then $P(s) = 0$, and if large enough, then $P(s) = 1$. In a statement like that, using either $<$ or $\leq$ makes no difference usually, but the default rule is to use $\leq$, and in that case $P(s)$ as defined above is called a cumulative probability function. We note that $P()$ is a heavily overloaded symbol: if used in this way, it can appear in many places in the text, to be interpreted according to context and arguments.

Random variables are other overloaded symbols, usually written with a capital letter, often X, Y, or Z. These symbols are each associated with a cumulative probability functions written, for example, as $F_X(q)$, meaning $F_X(q) = P(X \leq q)$. When, as in the example above, $X = |\mathbf{r}_1 - \mathbf{r}_2|$, it could be that F_X is discontinuous, but it is always nondecreasing in the range [0–1]. If it is discontinuous, it is so only at distinct values. In any case, there is also a probability density function, f_X, associated with X. If $F_X(s)$ has a discontinuity at $s = a$, then $f_X(s)$ is considered to contain a Dirac distribution $\delta(s - a)$. If s can only have discrete values, the same notation is used for the probability values, without bothering with any δ distribution:

$$P(X \leq x) = F_X(x),$$

$$(\text{if discrete}) = \sum_{t \leq x} f_X(x),$$

$$(\text{if continuous}) = \int_{t \leq x} f_X(x) \, dt.$$

This formalism can be extended to higher dimensionality, and to composite event spaces with both, say, integer values, real ranges, vector quantities, and, for example, angular distributions. These are treated analogously, using multivariate integral formulae. In that case, conversion between use of spherical polar coordinates and Cartesian coordinates will involve a Jacobian factor in the distribution functions.

In the above, the probability was assumed to be the event that some statement is true. One usually says that the event "has occurred." The event can also be regarded as a set, and one can then use the basic set or subset relations, for example, in formulae like

$$P(A \cup B) = P(A) + P(B) - P(A \cap B),$$

where $A \cup B$ is the set union, that is, the event that A or B has occurred, and $A \cap B$ is the set intersection, the event that both A and B has occurred.

There are many theorems for computing probabilities. As a useful example, given a number of events $A_1, A_2, \ldots$, one may want the probability that precisely k of these events occur. Compute the sums $S_1, S_2, \ldots$ as

$$S_1 = \sum_i P(A_i), \quad S_2 = \sum_{i<j} P(A_i \cap A_j), \quad S_3 = \sum_{i<j<k} P(A_i \cap A_j \cap A_k), \ldots$$

Then

$$P(\text{Exactly } k \text{ events}) = S_k - \binom{k+1}{k} S_{k+1} + \binom{k+2}{k} S_{k+2} - \cdots$$

(where $\binom{a}{b}$ stands for the usual binomial coefficients).

Conditional probabilities, such as the probability that A occurs, given that B occurs, is

$$P(A|B) = \frac{P(A \cap B)}{P(B)}, \quad \text{and similarly } P(B|A) = \frac{P(A \cap B)}{P(A)}.$$

Finally, there are certain statistical functions of the random variables: The simplest one, the expectation value of X, is written $E[X]$ or μ_X:

$$\text{Expectation,} \quad E[X] = \sum_{x \in A} x f_X(x) \qquad \text{(discrete)},$$

$$E[X] = \int_{x \in A} x f_X(x) \, dx \qquad \text{(continuous)},$$

$$\text{Variance,} \quad \text{Var}[X] = E[(X - \mu_X)^2]$$

$$\text{Covariance,} \quad \text{Cov}[X, Y] = E[(X - \mu_X)(Y - \mu_Y)].$$

Covariances are generally used to obtain numerical measures of correlation. If two variables have covariance 0, then they are uncorrelated. It should be noted that this does not mean that they are independent!

2.5 DENSITY FUNCTIONS FOR PARTICLES

We now assume that the state of a particle system is fully described by the positions and momenta, at any given time. If the forces acting in the system are known, then this is a deterministic system. Usually, we want a probabilistic description, because the full description is too elaborate, perhaps even infinite or open to random external influence. The principles are simple if we assume there exists a distribution function $\rho(x_1, x_2, \ldots, x_N)$, where $N = 6n$, n is the number of particles, and the variables are simply the Cartesian coordinates of positions and momenta. Suppose we are interested only in $f_1(\mathbf{x}), f_2(\mathbf{x})$—two real variables, which are functions of the N original variables. Now any statistics in the two new variables can be obtained from a reduced density function,

$$\rho'(y_1, y_2) = \int \delta(y_1 - f_1(\mathbf{x})) \delta(y_2 - f_1(\mathbf{x})) \rho(\mathbf{x}) \, dx_1 \cdots dx_N.$$

If the new variables are chosen in some way relevant to the problem at hand, then the equations of motion of the original description can be replaced by stochastic differential equations for ρ', valid at least for shorter time (a so-called Liouville equation). If the new variables are identical to the original variables x_1 and x_2, then we get especially simple equations,

$$\rho'(x_1, x_2) = \int \rho(\mathbf{x}) \, dx_3 \cdots dx_N$$

by "integrating away" the unwanted variables. Of course, this is an example of principle. Two variables are probably too few to be really useful.

2.6   WAVE FUNCTIONS AND DENSITY FUNCTIONS

If the wave function Ψ is known, the density function is the square of its complex modulus, or absolute value. Disregarding spin, it is

$$\rho(\mathbf{r}_1, \ldots, \mathbf{r}_n) = |\Psi(\mathbf{r}_1, \ldots, \mathbf{r}_n)|^2.$$

But this is just the configuration density, since the momentum variables are missing. Actually, the information about the momentum was erased when taking the absolute value. This becomes obvious if we do a $3n$-dimensional Fourier transform,

$$\tilde{\Psi}(\mathbf{p}_1, \ldots, \mathbf{p}_n) = (2\pi)^{-3n/2} \int \cdots \int \Psi(\mathbf{r}_1, \ldots, \mathbf{r}_n) \exp\left(-i \sum_{j=1}^{n} \mathbf{r}_j \cdot \mathbf{p}_j\right) d\mathbf{r}_1^3 \cdots d\mathbf{r}_n^3$$

since it can be shown that this is essentially a wave function using momentum variables instead of position variables:

$$\rho(\mathbf{p}_1, \ldots, \mathbf{p}_n) = |\tilde{\Psi}(\mathbf{p}_1, \ldots, \mathbf{p}_n)|^2.$$

In fact, the classical complete set of positions and momenta can be replaced by any other set of coordinates in $6n$-dimensional phase space, provided the coordinate transformation is one-to-one. Thus, a distribution of either position or momenta can be obtained from a wave function. In fact, a distribution can be obtained for any "complete set of observables," which is in principle a set of functions of classical position and momenta, expressed as the corresponding quantum mechanical operators, provided that these observables all commute with each other. This implies that there are wave functions that are simultaneously eigenfunctions to all the observables. The set should also be complete, meaning that each such eigenfunction describes a unique state, that is, they are unique apart from scaling.

There are a few complications involved. Two of these are notable: First of all, there are perfectly useful observables that have both discrete values and continuous ranges of values in their spectrum. Sums over discrete eigenvalues correspond to integrals over continuous ranges. Typically, we would sum over the spin eigenvalues, and integrate over position coordinates, but it is perfectly possible, for example, to use sums over angular eigenvalues of a subsystem, in which case a wave function would be decomposed into products of subsystem functions multiplied with functions of the remaining observables. As noted earlier, we may write this just as a sum or just as an integral, to be properly interpreted by case.

The second one is that electrons, as well as other fundamental particles, are indistinguishable. This means that, even if we write the wave function as depending on a number of individual positions (and spins), an observable in an electronic wave function must be a symmetric function involving all the particles. However, as noted elsewhere, this issue is automatically solved if a second quantization formalism is used.

Disregarding the complications, we note that there is also a great simplification, as soon as the observables relate to only a few particles at a time. The wave function above depends on all the coordinates of the complete system, and we are certainly not going to reexpress it into the relevant observables for each problem at hand. That would imply that, in order to compute an expectation value of the dipole operator in the z direction, say, we would have to transform the coordinate system, using $q_1 = z_1 + \cdots + z_n$ as one variable (and all other reexpressed to get an independent set of coordinates). Instead, we note that covariances, distribution functions over commuting variables, etc. can all be expressed in terms of probabilities (or probability densities), each of which is a quantum-mechanical expectation value, which can be computed as an integral. Thus, a joint probability distribution for m commuting observables

$$A_j(q_1, \ldots, q_n, p_1, \ldots, p_n), j = 1, \ldots, m$$

is computed by associating them with operators. If the variable q_j in the wave function is the Cartesian coordinate y, then $\hat{q}_j = y, \hat{p}_j = -i\partial/\partial y$. For a variable such as the total momentum, we would use $\sum_k - i\nabla_k$, and so on. When $\hat{A}_j$ are the corresponding operators, which do not have to commute with the arguments of Ψ, the expectation values and the covariances of the variables are

$$\langle A_j \rangle = \langle \Psi | \hat{A}_j | \Psi \rangle,$$

$$\langle A_j A_k \rangle = \langle \Psi | \hat{A}_j \hat{A}_k | \Psi \rangle,$$

$$\mathrm{Cov}(A_j A_k) = \langle A_j A_k \rangle - \langle A_j \rangle \langle A_k \rangle. \tag{2.9}$$

The variables do not have to be independent; it is quite common to compute, for example, expectations of several powers of some variables (called "moments").

Some probabilities or probability distributions where individual values of some property are singled out (rather than sums, like for covariances) cannot be done so easily, since the proper observables are then projection operators for the eigenfunctions of the observable, and not the observable itself. For a distribution over the distance $r_{12} = |\mathbf{r}_1 - \mathbf{r}_2|$, we may change the overlap integral to integrate only over the subspace where $r_{12} \leq s$, probably by using polar coordinates for the vector $\mathbf{r}_1 - \mathbf{r}_2$.

But the most flexible method is to leave the n-electron wave functions and use instead the second quantization formalism and reduced density matrices.

2.7 DENSITY MATRICES

The density matrix of a quantum state represents all information that is available about a stochastic mixture of pure states; the pure states are those that can be represented by a wave function. For a pure state, the density matrix is simply $\hat{\Gamma} = |\Psi\rangle\langle\Psi|$. It is used as a formal operator, or as an integral operator with an integration "kernel" $\Gamma(\mathbf{x}, \mathbf{x}')$, or as a matrix $\Gamma_{\mu,\nu}$, but these are all called "density matrix" and are just different representations of the same object.

Using a set of orthonormal wave functions $\Psi_\mu(\mathbf{x})$, it can thus take different forms:

$$\hat{\Gamma} = \sum_{\mu,\nu} \Gamma_{\mu,\nu} |\Psi_\mu\rangle\langle\Psi_\nu|, \tag{2.10}$$

$$\Gamma(\mathbf{x}, \mathbf{x}') = \sum_{\mu,\nu} \Gamma_{\mu,\nu} \Psi_\mu(\mathbf{x})\Psi_\nu(\mathbf{x}'). \tag{2.11}$$

One usage is when these wave functions are eigenfunctions to the Hamiltonian with eigenvalues E_μ, the matrix is diagonal with $\Gamma_{\mu,\nu} = P_\mu \delta_{\mu,\nu}$, and the values P_μ are interpreted as probabilities. These probabilities can be proportional to Boltzmann factors, $P_\mu = \exp(-E_\mu/(k_B T))$, and then the density matrix represents a canonical ensemble.

The density matrix thus has much more information than the wave functions, and when we (as is the usual case) are interested only in a few electronic states, it is not of much practical use. However, it is the top instance of a sequence of "reduced density matrices" of great importance. These have some useful properties in common.

The example above is only one special case of a density matrix. In general, a Hilbert space operator is called a density matrix or a density operator, if it is a Hermitian operator with nonnegative eigenvalues, whose sum is 1.

For any observable, let Φ be a normalized eigenfunction corresponding to the eigenvalue λ. Then the probability that this eigenvalue is "observed" in the system is obtained by "sampling" the density matrix using Φ in the "sandwich" formula:

$$P[\Phi] = \langle\Phi|\hat{\Gamma}|\Phi\rangle$$

(If the eigenvalue is continuous, we obtain instead a probability density.)

We observe that for a pure state, such a probability has the usual expression in terms of the wave function.

The same properties hold true (almost) for the reduced density matrices. These are obtained by integrating away all but a small subset of the variables of the wave function or density matrix. The usual one-electron and two-electron reduced density matrices are, respectively,

$$\gamma^{(1)}(y, z) = \int \cdots \int \Psi(y, x_2, \ldots, x_n)\Psi^*(z, x_2, \ldots, x_n) \, dx_2 \cdots dx_n, \tag{2.12}$$

$$\gamma^{(2)}(y, y'; z, z') = \frac{1}{2} \int \cdots \int \Psi(y, y', x_3, \ldots, x_n)\Psi^*(z, z', x_3, \ldots, x_n) \, dx_3 \cdots dx_n.$$

For an n-electron wave function, the sum of the eigenvalues is not one but n and $(n(n+1))/2$ for the one- and the two-electron reduced density matrix, respectively.

While the density matrix has probabilities as eigenvalues, the reduced matrices have *natural occupation numbers* and *natural pair occupation numbers*, for the one- and two-electron matrix, respectively. The eigenfunctions are called *natural spin-orbitals* and *natural geminals*, in the two cases.

It is common to use orbitals that are divided into two sets, one for α and one for β spin. Sorting them with α spin first, the matrix is naturally divided up into four submatrices:

$$\mathbf{D} = \begin{pmatrix} \mathbf{D}^{\alpha\alpha} & \mathbf{D}^{\alpha\beta} \\ \mathbf{D}^{\beta\alpha} & \mathbf{D}^{\beta\beta} \end{pmatrix}.$$

Assuming that the spin projection M_S is a "good quantum number," $\mathbf{D}^{\alpha\beta}$ and $\mathbf{D}^{\beta\alpha}$ are both zero. The diagonal values in the two remaining blocks are nonnegative, and they sum up to the expectation numbers of α and β electrons:

$$\sum_k D_{kk}^{\alpha\alpha} = n_\alpha \qquad \sum_k D_{kk}^{\beta\beta} = n_\beta.$$

If furthermore the wave function is an eigenfunction of both $\hat{S}^2$ and $\hat{S}_z$, and if $\mathbf{D}^{\alpha\alpha}$ and $\mathbf{D}^{\beta\beta}$ both use the same basis of spatial orbitals, then the whole set of related wave functions with the same $\hat{S}^2$ and different spin projection M_S, that is, the invariant set of states that is closed under spin rotation, have density matrix components that can be related using the so-called Wigner–Eckart rules. The sum of the α and β components is independent of M_S, and is called the *spin summed* or *spin free* density matrix. Its eigenvalues are between 0 and 2, called *natural occupation numbers*, and the eigenvectors represent *natural orbitals*.

One important aspect of the natural orbitals is the optimality property. If a selection of orbitals are used to form an approximate CI expansion, the largest overlap with the approximated wave function is obtained if its natural orbitals with highest occupation numbers are used [2].

It would be very useful to find some parametrization such that natural geminal wave functions could be constructed directly for an N-electron system with $N > 2$, but no practical way has been found except to start with an N-electron wave function. A two-electron pure state can be exactly reconstructed from its one-electron density matrix, and it is known that any Hermitian matrix with eigenvalues between 0 and 1 can be the one-electron density matrix of a physically realizable system. It could represent a mixed state, of course. It can represent a wave function, if the occupations add up to an exact integer. In particular, if all occupations are either 0 or 1, then the matrix represents a single Slater determinant, and it is then also a projection matrix. And, if it adds up to two and is known to represent a pure state, then this state can be exactly reconstructed (apart from an ambiguity of certain phase choices).

A pure state can be represented simply by the wave function itself. A typical other description is in terms of SDs, and the density matrix is then generally nondiagonal even for pure states.

As a simple example, take the diatomic molecule PN, which has a stable $^1\Sigma^+$ ground state. For orbitals, we have used the natural active orbitals of a small CASSCF;

these were the inactive molecular orbitals $1\sigma, \ldots, 6\sigma$ and $1\pi_x, 1\pi_y, 2\pi_x$, and $2\pi_y$, and the active orbitals were the $7\sigma, 8\sigma$ and $2\pi_x, 2\pi_y, 3\pi_x, 3\pi_y$, with six active electrons. The inactive orbitals remain atom-like at all distances. However, the active configuration is Hartree–Fock-like at shorter distances, that is, $\approx 7\sigma^2\, 2\pi^4$, and goes smoothly over to be $\approx 7\sigma^1\, 8\sigma^1\, 2\pi^2\, 3\pi^2$ at dissociation. Since at all distances, the state remains $^1\Sigma^+$ that is nondegenerate, it is always a pure state; the state operator is simply $\hat{\Gamma} = |\Psi\rangle\langle\Psi|$ and gives the same information as the wave function itself. The density matrix is zero apart from the single value 1 on the diagonal, if the basis consists of the eigenstates of the Hamiltonian.

If instead the basis was chosen to be spin-coupled configuration states constructed from the natural orbitals, the density matrix would be almost the same at the equilibrium distance, but at larger distances it would instead gradually become filled with numerous values. At 5 Å, the diagonal contains 16 nonzero values ranging from about 0.02 to 0.08; these values are the squares of the CI coefficients. Also, 240 nondiagonal elements would be nonzero with similar magnitudes but varying sign. The matrix still has only one nonzero eigenvalue.

This means that, although this is a pure state, it cannot be represented by a single configuration function, but 16 such functions are needed in order to represent the state even qualitatively correct. At large distance, a broken-symmetry Unrestricted HF calculation can give a passable approximation to the energy, but is then not a pure spin state. Being a single-determinant wave function, however, this is a pure state but with mixed spin.

A less conventional kind of reduced density matrices can be used to obtain description of quantum states of a smaller part of a molecule. In this example, we can for instance integrate out the spatial and spin variables associated with orbitals located on the nitrogen atom (say); at dissociation, this results in a state operator for the phosphorus atom, describing a *mixed* state—a statistical mixture of the four spin components of the $^4S^o$ state of the phosphorus ground state. A similar calculation for the nitrogen atom is obtained by integrating out the phosphorus orbitals; this is again a statistical mixture of nitrogen $^4S^o$ states. One concludes that when part of a larger system is considered in isolation, then even if the larger system has a well-defined wave function, the part under consideration is not described by a wave function but by a mixture of wave functions, and one has to use a density matrix for its accurate description.

Any matrix can be written as a so-called singular value decomposition which allows it to be described compactly in terms of a (possibly small) number of positive numbers and vectors; in the case of Hermitian matrices with positive eigenvalues, this is simply the eigenvalues and eigenvectors of the matrix. If this is done with an unconventional density matrix like the ones for P and N above, then their eigenvectors describe the minimal number of "natural" atomic states that are needed in order to describe PN with a so-called Generalized Valence Bond method. These atomic states are each associated with a singular value, showing its importance, or loosely speaking with which percentage the combination of the corresponding pair of atomic states contribute to the wave function. In this case, this would also include states with different number of electrons, showing, the "percentage" of charge transfer states, for example, P^-N^+, that contribute to the wave function.

2.8   REFERENCES

[1] Feller DF, Ruedenberg K. Systematic approach to extended even-tempered orbital bases for atomic and molecular calculations. Theor Chim Acta 1979;52:231–251.

[2] Davidson ER. Reduced Density Matrices in Quantum Chemistry. New York: Academic Press; 1976.

3

MOLECULAR ORBITAL THEORY

With the introduction of quantum mechanics, we got tools to solve the motion of the particles that build up molecules. In principle, we can compute all of chemistry by solving the Schrödinger equation which for one particle in an external field, represented by the potential energy $V(x, y, z)$, is written as

$$-\frac{\hbar^2}{2m}\left(\frac{\partial^2 \Psi}{\partial x^2} + \frac{\partial^2 \Psi}{\partial y^2} + \frac{\partial^2 \Psi}{\partial z^2}\right) + V(x, y, z)\Psi = E\Psi. \tag{3.1}$$

This is the time-independent Schrödinger equation that describes stationary states that do not evolve with time. This can be written more compactly as

$$\hat{H}\Psi = E\Psi, \tag{3.2}$$

where we introduce the Hamiltonian operator

$$\hat{H} = -\frac{\hbar^2}{2m}\left(\frac{\partial^2}{\partial x^2} + \frac{\partial^2}{\partial y^2} + \frac{\partial^2}{\partial z^2}\right) + V(x, y, z), \tag{3.3}$$

$$= -\frac{\hbar^2}{2m}\nabla^2 + V(x, y, z). \tag{3.4}$$

The Schrödinger equation can be extended to a system of more than one particle, for example, for the hydrogen atom we have

$$-\frac{\hbar^2}{2m_e}\nabla_e^2\Psi(r_e, r_p) - \frac{\hbar^2}{2m_p}\nabla_p^2\Psi(r_e, r_p) - \frac{e^2}{4\pi\epsilon_0 r}\Psi(r_e, r_p) = E\Psi(r_e, r_p), \tag{3.5}$$

Multiconfigurational Quantum Chemistry, First Edition.
Björn O. Roos, Roland Lindh, Per Åke Malmqvist, Valera Veryazov, and Per-Olof Widmark.
© 2016 John Wiley & Sons, Inc. Published 2016 by John Wiley & Sons, Inc.

where the first term is the kinetic energy of the electron, the second term is the kinetic energy of the proton, and the third term is the electrostatic interaction between the two particles. Solving this equation for the hydrogen atom leads us to a one-electron wave function, which we usually refer to as an orbital, and is the basis for atomic and molecular orbital theory.

3.1 ATOMIC ORBITALS

When we describe the electronic structure of atoms, we refer to atomic orbitals, that is, solutions to a one-electron Schrödinger equation. The motion of each electron is described by one of these orbitals, and the total wave function of the atom is described by the product of such one-electron wave functions. The orbital that describes the motion of one electron is said to be populated, or occupied, with an electron while the remaining orbitals that are not used in describing the motion of an electron are said to be unpopulated or unoccupied .

The logical place to start is by studying the hydrogen atom that indeed only has one electron and is described properly by one occupied orbital.

3.1.1 The Hydrogen Atom

The hydrogen atom is the smallest and simplest system of chemical interest. It consists of two particles: an electron and a proton. We can use the fact that the mass of the proton is much higher than that of the electron and make the approximation that the collective motion of the atom coincides with the nucleus. The motion of the center of mass can be described by a free particle and the relative motion of the two particles remains to be solved. Doing this we get effectively a one-particle problem to solve, the motion of an electron of mass m_e orbiting around a stationary nucleus.

On the other hand, it is simple to make a separation of the collective motion of the atom and the relative motion of the two particles. Again we effectively obtain a one-particle problem to solve, the motion of an electron around the common center of mass, but now we must use the reduced mass of the system, $\mu = \frac{m_e m_p}{m_e + m_p} \approx m_e$, where m_e is the electron mass and m_p is the proton mass. We have the Hamiltonian

$$\hat{H} = -\frac{\hbar^2}{2\mu}\nabla^2 - \frac{e^2}{4\pi\epsilon_0 r} = -\frac{1}{2\mu}\nabla^2 - \frac{1}{r} \tag{3.6}$$

for this system, where r is the distance between the two particles. The Schrödinger equation for this system can be solved analytically, and is most conveniently represented in spherical polar coordinates. The solutions then take the form

$$\Psi_{nlm_l}(r, \theta, \varphi) = R_{nl}(r)Y_{lm_l}(\theta, \varphi), \tag{3.7}$$

where $Y_{lm_l}(\theta, \varphi)$ are the well-known spherical harmonics and $R_{nl}(r)$ is given by

$$R_{nl}(r) = N_{nl}\rho^l L_{n+1}^{2l+1}(\rho)e^{-\rho/2}; \quad \rho = 2r/na_0; \quad a_0 = \frac{4\pi\epsilon_0\hbar^2}{m_e e^2} = 1, \qquad (3.8)$$

where a_0 is the so-called Bohr radius, $R_{nl}(r)$ determines the radial behavior, and $Y_{lm_l}(\theta, \varphi)$ determines the angular behavior. $L_{n+1}^{2l+1}(\rho)$ are the associated Laguerre polynomials.

The wave functions depend on three quantum numbers n, l, and m_l, which have different physical meaning.

- n is the so-called principal quantum number that gives the energy (E_n) of orbital n, also specifies the sum of the number of radial nodes and angular nodes ($n - 1$ nodes).
- l specifies the magnitude of the angular momentum of the electron around the center of mass, usually referred to as the orbital angular momentum quantum number with the magnitude being $\sqrt{l(l+1)}\hbar$. Also the number of angular nodes (l nodes) are given by l.
- m_l is the magnetic quantum number that specifies the z component of the orbital angular momentum with the value $m_l\hbar$.

These quantum numbers cannot take arbitrary values; there are some restrictions:

- n takes the values $1, 2, 3, \ldots, \infty$ with the corresponding energy $E_n = -\frac{Z^2\mu}{2n^2}$.
- l takes the values $0, 1, 2, \ldots, n-1$ that restricts the possible magnitudes of the orbital angular momenta for a given principal quantum number.
- m_l takes the values $0, \pm1, \ldots, \pm l$. Not surprising, the z component of the orbital momentum can not be larger than the magnitude of the orbital angular momentum.

The spherical harmonics are complex valued functions, an inconvenience that can be remedied. In the absence of a magnetic field, the energy of an orbital is independent on m_l that allows us to make linear combinations of Y_{lm_l} with fixed l and varying m_l and can do so that we only obtain real-valued functions. For example, $(Y_{1,1} + Y_{1,-1}) \propto x/r$, which can be combined with the radial part, $R_{n1}(r)$, to yield a p_x orbital. In a similar manner, we can obtain p_y and p_z orbitals. Forming linear combinations of Y_{2,m_l} will yield d-orbitals, Y_{3,m_l} f-orbitals, etc.

For historical reasons, we call orbitals with angular momentum $l = 0$ s orbitals, $l = 1$ p orbitals, $l = 2$ d orbitals, $l = 3$ f orbitals, etc.

For $n = 1$, the only value for l is 0 (zero), and since $n = 1$ we name this orbital 1s.

For $n = 2$, one possibility is that $l = 0$ and we call this orbital 2s, similarly to 1s. Another possibility is that $l = 1$, and in a similar manner we name these orbitals 2p. Here we have three degenerate components that can be represented by np_x, np_y, and np_z.

For $n = 3$, we get 3s and 3p as for $n = 2$ and in addition 3d orbitals. For $n = 4$, we also get 4s, 4p, 4d as well as 4f orbitals, etc.

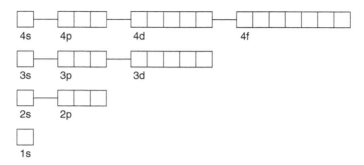

Figure 3.1   Shells and subshell of atoms.

All orbitals belonging to a given principal quantum number constitute a shell, while a subshell is given by the principal quantum number and the orbital angular momentum quantum number. For example, 3s, 3p, and 3d jointly form a shell while 3p is a subshell (see Figure 3.1).

3.1.2   The Helium Atom

We were able to solve the hydrogen atom case and obtained wave functions describing the motion of the electron around the nucleus. If we now turn to the helium case, we have three particles: a nucleus and two electrons. Assuming the approximation of a heavy nucleus, the collective motion of the atom can be regarded as the motion of the nucleus. Again the collective motion is described by the free particle case and is ignored leaving us with the positions of the two electrons relative to the nucleus, that is, we have six coordinates describing our system. We can conveniently, and quite arbitrarily, place the nucleus at the origin of our coordinate system.

The wave function describing the electronic motion is now dependent on the positions of two electron, and further rewriting of the Hamiltonian in such a way that we can separate the variables is not possible. This equation cannot be solved analytically and we need to make some approximations. Interpreting a function in six dimensions is much more difficult than interpreting a function in three dimensions.

At this point, we can make the apparently crude approximation of ignoring the electron–electron interaction, and for such a system, we have a Hamiltonian that can be separated into the sum of two Hamiltonians, one for each electron. In this approximation, the motion of one electron is independent of the other. For such a system, the wave function can simply be written as a product of the wave function for each electron:

$$\Psi(r_1, r_2) = \psi_1(r_1)\psi_2(r_2). \tag{3.9}$$

These one-electron wave functions are usually referred to as orbitals, so the wave function is here a product of two orbitals.

Orbitals is a very useful notion; we can fairly easily make a mental picture of how a function behaves in three dimensions but not six. Further, we have made it possible to interpret properties of a system in terms of the properties of each individual orbital.

On the other hand, we can, without making any approximations to the Hamiltonian, use an approximate wave function in the form

$$\Psi(r_1, r_2) \approx \psi_1(r_1)\psi_2(r_2), \tag{3.10}$$

the so-called orbital approximation. With such a wave function, we can use the variational principle to obtain the best possible wave function, in the sense of yielding the lowest possible energy, that is, minimize $\langle \Psi | \hat{H} | \Psi \rangle$ with the constraint $\langle \Psi | \Psi \rangle = 1$. This leads to equations where each electron moves in the potential created by the nucleus and the average density of the other electron. This is the basis for Hartree–Fock theory covered in Chapter 4. Solving these effective one-electron equations, we get the same set of orbitals as for the hydrogen atom, 1s, 2s, 2p, 3s, etc. It should be noted, however, that the radial shape is quite different; for example, the 1s of U^{91+} have an exponential decay proportional to e^{-92r/a_0} compared to e^{-r/a_0} for hydrogen. Adding the remaining 91 electrons to the uranium atom does not change this exponential decay drastically.

Making the approximation that the wave function is a product of orbitals is the fundamental orbital approximation and it is the basis for most of the quantum chemistry today.

There is one fundamental flaw with the form of Eq. 3.10, namely the permutation symmetry. The square norm of the wave function $|\Psi(r_1, r_2)|^2$ is the probability density of finding particles 1 and 2 at given positions. Two electrons are indistinguishable particles and the physics should not be affected by our numbering; thus, we expect that $|\Psi(r_1, r_2)|^2 = |\Psi(r_2, r_1)|^2$. This can be accomplished by $\Psi(r_2, r_1) = \pm\Psi(r_1, r_2)$ where the minus sign is correct for particles with half integral spin, such as electrons, protons, and neutrons. This is the Pauli principle for fermions. To fulfill this, we must write the product as

$$\Psi(r_1, r_2) \approx \frac{1}{\sqrt{2}}(\psi_1(r_1)\psi_2(r_2) - \psi_2(r_1)\psi_1(r_2)) \tag{3.11}$$

except for an arbitrary phase factor.

Electrons do have an intrinsic angular momentum we call spin with the magnitude $\sqrt{s(s+1)}\hbar$ where $s = \frac{1}{2}$ and the z component $m_s\hbar$, where $m_s = \pm\frac{1}{2}$. Electrons with $m_s = \frac{1}{2}$ are usually called α electrons and electrons with $m_s = -\frac{1}{2}$ are usually called β electrons. To fully specify the wave function for one electron, we need to specify both the spatial part as well as the spin part. For the ith electron in the kth spin orbital, we can conveniently write $\psi_k(i) = \varphi_k(i)\sigma_k(i)$, where ψ_k is the spin orbital, φ_k is the spatial part, and σ_k is the spin part.

The lowest possible energy for the helium atom is when both electrons are in the orbital with the lowest energy, namely the 1s orbital. Using a more compact notation, we can now write the wave function as

$$\Psi(1,2) \approx \frac{1}{\sqrt{2}}(\varphi_{1s}(1)\sigma_1(1)\varphi_{1s}(2)\sigma_2(2) - \varphi_{1s}(1)\sigma_2(1)\varphi_{1s}(2)\sigma_1(2)), \qquad (3.12)$$

$$\approx \varphi_{1s}(1)\varphi_{1s}(2)\frac{1}{\sqrt{2}}(\sigma_1(1)\sigma_2(2) - \sigma_2(1)\sigma_1(2)), \qquad (3.13)$$

where σ_1 and σ_2 are the spins for the two particles. It is obvious that $\sigma_1 \neq \sigma_2$ and we can quite arbitrarily write $\sigma_1 = \alpha$ and $\sigma_2 = \beta$. Now we can write the wave function as a product of a spatial wave function, Ψ_{space}, and a spin wave function Ψ_{spin} written as

$$\Psi_{space} = \varphi_{1s}(1)\varphi_{1s}(2), \qquad (3.14)$$

$$= \varphi_{1s}\varphi_{1s}, \qquad (3.15)$$

$$\Psi_{spin} = \frac{1}{\sqrt{2}}(\alpha(1)\beta(2) - \beta(1)\alpha(2)), \qquad (3.16)$$

$$= \frac{1}{\sqrt{2}}(\alpha\beta - \beta\alpha), \qquad (3.17)$$

where we have used implied particle numbering according to position in the last short-hand notation.

3.1.3 Many Electron Atoms

The helium atom in its ground state has two electrons in the 1s orbital. If we go one step further and consider the lithium atom with three electrons, it might be tempting to try to put all three electrons in the 1s orbital. This would give us a $\Psi_{space}(1,2,3)$ that is symmetric in interchange of any two indices, that is $\Psi_{space}(2,1,3) = \Psi_{space}(1,2,3)$, for example, and we would need a $\Psi_{spin}(1,2,3)$ that is antisymmetric in interchange of any two indices. It is not possible to make a spin wave function that is antisymmetric in interchange of any two indices. This leads us to the Pauli exclusion principle

There may not be more than two electrons in any given orbital. If an orbital contains two electrons they must have opposite spin.

This also leads us to the important aufbau (German for building up) principle that tells us how to add electrons as we move across the periodic table. We fill shells and subshells according to

1s, 2s, 2p, 3s, 3p, 4s, 3d, 4p, 5s, 4d, 5p, 6s, ...

which is regular up to 3p. Then we start filling up the 4s subshell rather than the 3d subshell.

For the hydrogen atom, we have degeneracy within a shell; the energy for 3s is the same as for 3d, for example. For any atom with two or more electrons, this degeneracy is lifted and the orbital energies increase as we go from one subshell to the next, for example, $\epsilon_{4s} < \epsilon_{4p} < \epsilon_{4d} < \epsilon_{4f}$. The reason for this is that electrons in high orbital angular momentum orbitals are more effectively screened by the inner electrons. When we get to the potassium atom in the periodic table, the orbital energy for 4s is actually lower than the orbital energy for 3d.

In atoms with more than one electron, the spin angular momentum adds up to give a total spin of the atom. If all subshells are completely filled or empty, the net angular momentum from the spin is zero. It is thus only from partially filled subshells we get a nonzero total spin. For the ground state of the atom, the electrons within a partially filled subshell spin align in the same direction as far as possible to give the largest total spin possible.

3.2 MOLECULAR ORBITALS

When constructing molecular orbitals, we may follow the same procedure as for atoms; compute the molecular orbitals for one electron orbiting around the nuclei and consider all one-electron wave functions, orbitals, that arise when we solve the Schrödinger equation. We can then populate the lowest molecular orbitals with electrons and again use the variational principle to determine the best wave function within the orbital approximation.

On the other hand, we may start with atom that has orbitals populated with electrons. We can then form molecular orbitals by combining the atomic orbitals into molecular orbitals. The latter approach more easily yields insight in the processes involved in chemical bonding, for example.

3.2.1 The Born–Oppenheimer Approximation

When we consider molecules, we have a number of electrons and a number of nuclei and we need to solve the Schrödinger equation for all these particles. There are good reasons to try to separate the motion of the electrons from the motion of the nuclei. The first and perhaps the most obvious reason is that the complexity of the calculation becomes much reduced. Another, and more important, reason is that we introduce the concept of potential energy surfaces, PES, this way.

The mass of the proton is about 1836 times that of the electron. According to the equipartition theorem, the kinetic energy is distributed evenly among all particles in an ensemble of particles, in an average sense. Considering this, an electron should on the average move about $43 \approx \sqrt{1836}$ times faster than a proton in a molecule, and for all other nuclei the difference becomes even larger.

It seems safe to assume that the (fast) electrons have plenty of time to rearrange their motion, whenever the (slow) nuclei change position.

The Hamiltonian for a molecule can be written as

$$\hat{H} = \hat{T}_n + \hat{T}_e + \hat{V}_{ne} + \hat{V}_{nn} + \hat{V}_{ee}, \tag{3.18}$$

where the first term is the kinetic energy for the nuclei, the second term is the kinetic energy for the electrons, the third term is the nuclei–electron attraction, the fourth term is the nuclei–nuclei repulsion, and the last term is the electron–electron repulsion. We can rewrite this as

$$\hat{H} = \hat{T}_n + \hat{H}_{el}, \tag{3.19}$$

where $\hat{H}_{el}$ contains all terms except the kinetic energy for the nuclei.

We can solve the equation

$$\hat{H}_{el}\Psi_{el}(r_e; r_n) = E_{el}(r_n)\Psi_{el}(r_e; r_n), \tag{3.20}$$

where the energy $E_{el}(r_n)$ depends on the position of the nuclei and the wave function Ψ_{el} depends parametrically on the position of the nuclei. This solution would be the exact description of the motion of the electrons in a molecule if the electrons would adjust their motion instantaneously when the positions of the nuclei change.

If this approximation were exact, we can write the wave function as $\Psi(r_e, r_n) = \Psi_{nuc}(r_n)\Psi_{el}(r_e; r_n)$ and finally solve for $\Psi_{nuc}(r_n)$ and we the get the equation

$$[\hat{T}_n + E_{el}(r_n)]\Psi_{nuc}(r_n) = E\Psi_{nuc}(r_n). \tag{3.21}$$

This is the so-called Born–Oppenheimer approximation that is usually a very good approximation but can fail miserably when large rearrangements of the electronic structure occur very abruptly with small changes in nuclear positions.

Apart from making calculation on molecular systems feasible, it introduces the concept of potential energy surface, PES, given by E_{el}. This is a powerful tool for analyzing and understanding reaction mechanisms.

3.2.2 The LCAO Method

When computing orbitals for molecules, it is convenient to start with atomic orbitals and use these to describe molecular orbitals. This approach is called linear combination of atomic orbitals or LCAO for short. It should be noted that LCAO in its basic form where we use the occupied atomic orbitals in our description yields approximate MOs.

The Hydrogen Molecule Ion   The simplest molecular system is H_2^+, which consists of three particles: an electron and two protons. Using the Born–Oppenheimer approximation, we only have to address the motion of one particle, the electron.

We start to look at this system where the distance between the two nuclei is very large so that we have no interaction. Without any further information, all we know is that we have the electron on one of the two nuclei but not which one. The probability of finding the electron in a 1s orbital on atom 1, φ_1, and the probability of finding the electron in a 1s orbital on atom 2, φ_2, is equal. This means that the probability density will look like $|\Psi|^2 = 0.5|\varphi_1|^2 + 0.5|\varphi_2|^2$, that is, a 50/50 chance of finding the electron on either atom. There are two ways of we can formulate a wave function

Figure 3.2   Two hydrogen atoms at large separation.

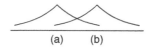

Figure 3.3   Two hydrogen atoms close to equilibrium.

that fulfills this condition, namely, $\Psi = \frac{1}{\sqrt{2}}(\varphi_1 \pm \varphi_2)$. At large separations, the sign has no effect on the physics, both solutions are eigenfunctions to $\hat{H}$, with the same energy. Figure 3.2 shows schematically the $\Psi = \frac{1}{\sqrt{2}}(\varphi_1 + \varphi_2)$ wave function for two hydrogen atoms at large separations.

What happens when the two nuclei approach each other and start to perturb the potential in each of the two potential wells? Obviously, the simple combination of atomic orbitals is not going to be the exact wave function for H_2^+ at intermediate distances.

Using the notion that molecules are really built up of atoms, we can *approximate* the molecular orbitals with Linear Combinations of Atomic Orbitals, which is commonly referred to by its acronym LCAO. This is a very useful and powerful concept that we can use to describe chemical bonding and reactions. Figure 3.3 shows the two atomic 1s orbitals of H_2 close to equilibrium.

Using the two atomic 1s orbitals on the two hydrogen nuclei as a basis for describing H_2^+, we can solve the Schrödinger equation within this subspace. For this purpose, let us express the wave function (orbital) as a linear combination of the two atomic orbitals, $\varphi_{\text{trial}} = c_1\varphi_1 + c_2\varphi_2$. Let us for simplicity assume that the orbitals (φ_i) and coefficients(c_i) are real valued and further that the phase of the atomic orbitals is such that $\varphi_i(\boldsymbol{r}) > 0$. The variation principle states that the expectation value for the energy of any wave function is higher than that for the exact ground state wave function. This can be expressed as

$$E_{\text{exact}} = \langle \varphi_{\text{exact}} | \hat{H} | \varphi_{\text{exact}} \rangle \leq E_{\text{trial}} = \langle \varphi_{\text{trial}} | \hat{H} | \varphi_{\text{trial}} \rangle \qquad (3.22)$$

with equality only if $\varphi_{\text{trial}} = \varphi_{\text{exact}}$.

We can minimize $\langle \varphi_{\text{trial}} | \hat{H} | \varphi_{\text{trial}} \rangle$ with the constraint $\langle \varphi_{\text{trial}} | \varphi_{\text{trial}} \rangle = 1$ to obtain solutions within this subspace. Let us first evaluate $\langle \varphi_{\text{trial}} | \hat{H} | \varphi_{\text{trial}} \rangle$ and $\langle \varphi_{\text{trial}} | \varphi_{\text{trial}} \rangle$.

$$\langle \varphi_{\text{trial}} | \hat{H} | \varphi_{\text{trial}} \rangle = |c_1|^2 \langle \varphi_1 | \hat{H} | \varphi_1 \rangle + c_1^* c_2 \langle \varphi_1 | \hat{H} | \varphi_2 \rangle$$
$$+ c_2^* c_1 \langle \varphi_2 | \hat{H} | \varphi_1 \rangle + |c_2|^2 \langle \varphi_2 | \hat{H} | \varphi_2 \rangle$$

$$= |c_1|^2 H_{11} + c_1^* c_2 H_{12} + c_2^* c_1 H_{21} + |c_2|^2 H_{22}$$
$$= c_1^2 H_{11} + c_1 c_2 (H_{12} + H_{21}) + c_2^2 H_{22}$$
$$= c_1^2 H_{11} + 2 c_1 c_2 H_{12} + c_2^2 H_{22}, \tag{3.23}$$

where $H_{nm} = \langle \varphi_n | \hat{H} | \varphi_m \rangle$ and

$$\langle \varphi_{\text{trial}} | \varphi_{\text{trial}} \rangle = |c_1|^2 \langle \varphi_1 | \hat{H} | \varphi_1 \rangle + c_1^* c_2 \langle \varphi_1 | \hat{H} | \varphi_2 \rangle$$
$$+ c_2^* c_1 \langle \varphi_2 | \hat{H} | \varphi_1 \rangle + |c_2|^2 \langle \varphi_2 | \hat{H} | \varphi_2 \rangle$$
$$= |c_1|^2 + c_1^* c_2 S + c_2^* c_1 S + |c_2|^2$$
$$= c_1^2 + 2 c_1 c_2 S + c_2^2$$
$$= 1, \tag{3.24}$$

where we have assumed that the atomic orbitals are normalized and $S = \langle \varphi_1 | \varphi_2 \rangle = \langle \varphi_2 | \varphi_1 \rangle$. Minimizing the energy now gives the equation

$$\begin{pmatrix} H_{11} & H_{12} \\ H_{11} & H_{22} \end{pmatrix} \begin{pmatrix} c_1 \\ c_2 \end{pmatrix} = E_{\text{trial}} \begin{pmatrix} 1 & S \\ S & 1 \end{pmatrix} \begin{pmatrix} c_1 \\ c_2 \end{pmatrix}. \tag{3.25}$$

The integrals $H_{11} = H_{22} = \alpha$ are usually referred to as the Coulomb integral and $H_{12} = H_{21} = \beta$ as the resonance integral.

To solve this equation for the energy, we can form the secular determinant

$$\begin{vmatrix} \alpha - E_{\text{trial}} & \beta - S E_{\text{trial}} \\ \beta - S E_{\text{trial}} & \alpha - E_{\text{trial}} \end{vmatrix} = 0, \tag{3.26}$$

which has the solutions

$$E_{\text{trial}} = \frac{\alpha \pm \beta}{1 \pm S}, \tag{3.27}$$

where the plus sign corresponds to $c_1 = c_2 = [2(1 + S)]^{-\frac{1}{2}}$ and the minus sign corresponds to $c_1 = -c_2 = [2(1 - S)]^{-\frac{1}{2}}$. The former has lower energy and corresponds to a bonding orbital, while the latter corresponds to an antibonding orbital.

Symmetry arguments would also have given us the solutions $c_1 = \pm c_2$, and we can classify these orbitals by their symmetry. In a diatomic molecule, the angular momentum of the electron is not a preserved quantity, but the z-component (assuming that the bond is along the z-axis) of the angular momentum is. An orbital with no angular moment around the z-axis is called a σ orbital. An orbital with $m_l = \pm 1$ is called a π orbital and an orbital with $m_l = \pm 2$ is called a δ orbital, in analogy to s, p, d, f, etc. for the atom. If we assume that the two nuclei are placed symmetrically around the origin, we can invert the coordinates and obtain the same molecule back. A solution to the Schrödinger equation for a system with inversion symmetry will be an even function or an odd function with respect to inverting the coordinates. If the function is even, it is labeled *gerade* (German for even) or *ungerade* (German for odd), and we use subscripts g and u: $\varphi_g(-x, -y, -z) = \varphi_g(x, y, z)$ and $\varphi_u(-x, -y, -z) = -\varphi_u(x, y, z)$. Thus, the bonding orbital is an orbital of σ_g symmetry while the antibonding orbital is of σ_u symmetry. They are the first orbital of their kind and are thus labeled $1\sigma_g$ and $1\sigma_u$, respectively.

It is relatively straightforward to demonstrate that the bonding orbital leads to a buildup of charge between the two atoms as compared to the added atomic densities. In the midpoint (r_m) between the two atoms, the amplitude of the two atomic orbitals are equal: $\varphi_1(r_m) = \varphi_2(r_m)$. With on the average half an electron on each atom the averaged atomic density is simply given by $\rho(r_m) = \frac{1}{2}(|\varphi_1(r_m)|^2 + |\varphi_2(r_m)|^2) = |\varphi_1(r_m)|^2 = \rho_1(r_m)$. The density from the bonding orbital is given by

$$
\begin{aligned}
\rho(r_m) &= \left| \frac{\varphi_1(r_m) + \varphi_1(r_m)}{\sqrt{2(1+S)}} \right|^2 \\
&= \frac{|\varphi_1(r_m)|^2 + \varphi_1^*(r_m)\varphi_2(r_m) + \varphi_2^*(r_m)\varphi_1(r_m) + |\varphi_2(r_m)|^2}{2(1+S)} \\
&= \frac{4\rho_1(r_m)}{2(1+S)} \\
&> \rho_1(r_m).
\end{aligned}
\tag{3.28}
$$

The inequality holds at all nonzero distances since S must be less than one.

Using the two 1s functions as basis functions, we can solve the Schrödinger equation exactly in the space defined by these basis functions. This gives us a simple tool to describe, for example, chemical bonding, and Figure 3.4 shows the bonding and antibonding potential curve for H_2^+. We can thus describe the chemical bond in a qualitative manner, but if we want results that are quantitatively correct, we need to be more sophisticated in the choice of the basis functions, which will be described in more detail in Chapter 6.

The Hydrogen Molecule After describing the H_2^+ ion, the next logical thing to describe is the H_2 molecule. H_2 has two electrons and we again use the orbital approximation to get an approximate wave function that is easy to handle. To describe the ground state, we occupy the lowest orbital, $1\sigma_g$, with two electrons of opposite spin giving us a singlet wave function. Within this basis set, there are three different configurations possible: $1\sigma_g^2$, $1\sigma_g 1\sigma_u$, and $1\sigma_u^2$. All three give rise to a singlet state but only the $1\sigma_g 1\sigma_u$ can give rise to a triplet state. Only the $1\sigma_g^2$ configuration give rise to a stable molecule and the approximate wave function is given by

$$
\Psi(1, 2) \approx \Psi_{MO}(1, 2) = 1\sigma_g(1)1\sigma_g(2)\frac{1}{\sqrt{2}}(\alpha(1)\beta(2) - \beta(1)\alpha(2)).
\tag{3.29}
$$

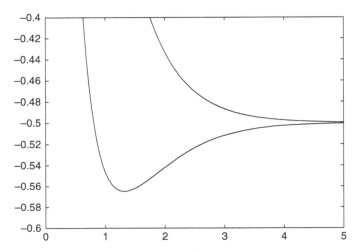

Figure 3.4   Potential curves of $1\sigma_g$ and $1\sigma_u$ of H_2^+. Bond distance and energy in atomic units.

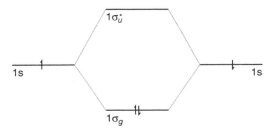

Figure 3.5   MO diagram for H_2 molecule.

MO diagrams are often used to represent the bonding in molecules, see Figure 3.5. To the left and the right in the diagram are the orbital energies drawn for the atomic orbitals; in this case, the two atomic 1s orbitals. In the middle, the orbital energies for the molecular orbitals are drawn; in this case, the $1\sigma_g$ and the $1\sigma_u$ orbitals. There are also lines connecting the atomic orbitals with the molecular orbitals. In this case, the connections are trivial; but in more complex cases, it may be very informative to indicate which atomic orbitals contribute to which molecular orbitals. In Figure 3.5, the electrons are also indicated with arrow up and arrow down indicating α and β spin, respectively.

3.2.3   The Helium Dimer

When we describe the bonding in He_2, we again start with the atomic 1s orbitals and form $1\sigma_g$ and $1\sigma_u$; but in this case, we have four electrons to distribute in the two molecular orbitals and we get the configuration $1\sigma_g^2 1\sigma_u^2$, that is, two bonding

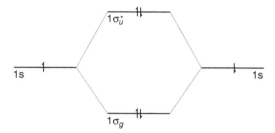

Figure 3.6   MO diagram for He_2 molecule.

and two antibonding electrons illustrated in Figure 3.6. Counting two bonding and two antibonding electrons suggest that there should be no bond formed; the effect of the antibonding orbitals canceling the effect of the bonding electrons. In fact, the antibonding effect is in general slightly larger than the effect of the bonding electrons; so He_2 is in fact repulsive, except for a very weak van der Waals minimum at about $3\,\text{Å}$.

3.2.4   The Lithium and Beryllium Dimers

When we come to the lithium dimer, the atoms we start with have a $1s^2 2s$ electron configuration. The 1s electrons are much more compact than the 2s electron, and any bonding that occurs only involves the 2s while 1s is inert. See Figure 3.7 for the radial distribution functions for 1s and 2s of Li. In Figure 3.8, we can see the MOs in Li_2 and the $1\sigma_g$–$1\sigma_u$ splitting is not visible at that scale, also indicating that the 1s orbitals do not contribute to the bonding. Normally, you would exclude the core orbitals in an MO-diagram since they do not contribute to the bonding.

Excluding the 1s contributions for Be_2, the resulting MO-diagram would look the same as for He_2 except that the proper labels for the MOs would be $2\sigma_g$ and $2\sigma_u$. Again we conclude that there is no bonding since we have as many bonding electrons as antibonding.

3.2.5   The B to Ne Dimers

Boron is the first atom to have p-electrons in the ground state and the molecular orbitals from the 2p orbitals alone is shown in Figure 3.9. From the two 2p components (one from each atom) with no angular momentum around the bond axis ($2p_z$) we form σ orbitals, $3\sigma_g$ and $3\sigma_u$, and the four components with $|m_l| = 1$ ($2p_x, 2p_y$) we form π orbitals, $1\pi_u$ and $1\pi_g$. The splitting of bonding and antibonding orbitals is closely related to the overlap between the atomic orbitals, and the p orbitals pointing in the direction of the bond have larger overlap than the two p orbitals that are perpendicular to the bond axis; hence, the splitting of the σ orbitals is larger than the splitting of the π orbitals.

When we form a molecule, the spherical symmetry of the atom is lost and in the case of a homonuclear diatomic molecule, we have the axial symmetry of the point

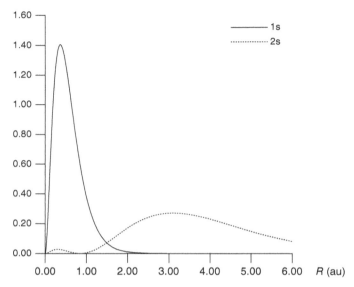

Figure 3.7   Radial distribution function $P(r) = r^2 R_{nl}(r)^2$ for 1s and 2s of lithium atom, atomic units.

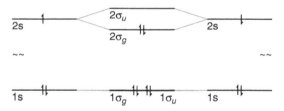

Figure 3.8   MO diagram for Li_2 molecule. (The energy difference between 1s and 2s levels is not shown with the correct scale.)

group $D_{\infty h}$. In this case, the symmetric combination of 2s $(2\sigma_g)$ and the symmetric combination of $2p_z$ $(3\sigma_g)$ will mix and leads to a stabilization of $2\sigma_g$ and a destabilization of $3\sigma_g$. The same will happen to the antibonding combinations; $2\sigma_u$ will be stabilized while $3\sigma_u$ will be destabilized. This is shown in Figure 3.10. The π orbitals are the first in their respective symmetry and there in no stabilization or destabilization due to mixing. As a consequence of this, the $3\sigma_g$ orbital will have slightly higher orbital energy than $1\pi_u$, and B_2 has a triplet ground state.

When we go to the right in the periodic system, the orbital energies of 2s and 2p become lower; they are more tightly bound for F than for B. Further the difference in energy becomes larger between 2s and 2p, and as a consequence the mixing of 2s and $2p_z$ becomes less important and the stabilization/destabilization is less pronounced. The MO diagram for F_2 is shown in Figure 3.11.

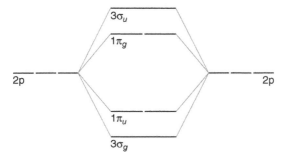

Figure 3.9   MO diagram for two 2p orbitals.

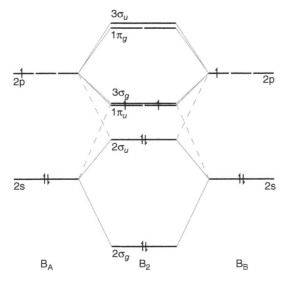

Figure 3.10   MO diagram for the B_2 molecule.

3.2.6   Heteronuclear Diatomic Molecules

When we consider heteronuclear diatomic molecules, much is the same as for homonuclear diatomic molecules, but a couple of things are different. The inversion symmetry does no longer exist; we cannot invert the coordinates of the molecule and get the same thing back. The labeling of orbitals would be 1σ, 2σ, 3σ, etc. without the gerade and ungerade notation.

Lack of inversion symmetry means that the mixing of atomic orbitals does not need to be equal for the two atoms. Further the orbital energies for the two atoms differ, and the difference indicates how the atomic orbitals mix when forming bonding and antibonding orbitals.

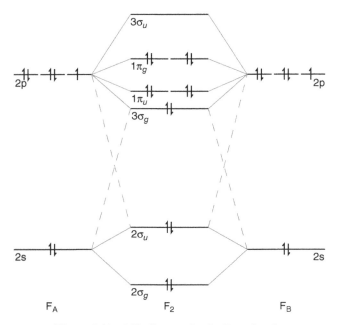

Figure 3.11 MO diagram for the F_2 molecule.

If we consider the HF molecule, there are two atomic orbitals of σ symmetry (if we ignore the 1s and 2s of F), the $2p_z$ of F and the 1s of H. The other valence orbitals are $2p_x$ and $2p_y$, which are π orbitals with no π orbitals on H to interact with and are nonbonding orbitals. We can express the bonding (and antibonding) orbital as a linear combination of the hydrogen 1s and fluorine $2p_z$; $\varphi = c_F\varphi_F + c_H\varphi_H$ in a similar manner to the H_2^+ case. We obtain the equation

$$\begin{pmatrix} \alpha_F & \beta \\ \beta & \alpha_H \end{pmatrix} \begin{pmatrix} c_F \\ c_H \end{pmatrix} = E \begin{pmatrix} 1 & S \\ S & 1 \end{pmatrix} \begin{pmatrix} c_F \\ c_H \end{pmatrix} \tag{3.30}$$

to solve. The equation now has two different coulomb integrals, α_F and α_H. If we solve this equation, we will obtain a bonding orbital and an antibonding orbital, where the bonding is mostly of F character and the antibonding mostly of H character, a result that is supported by the MO diagram in Figure 3.12. This bond is ionic and we almost have H^+F^-.

The CO molecule is isoelectronic with the N_2 molecule and not only exhibits similarities but also differences. The orbital energies for O is lower than those of C and further the difference between 2s and 2p is larger for oxygen, see Figure 3.13. Also 6 of the 10 valence electrons come from oxygen while 4 come from carbon, distributed into 4 bonding and 1 antibonding orbitals giving a triple bond.

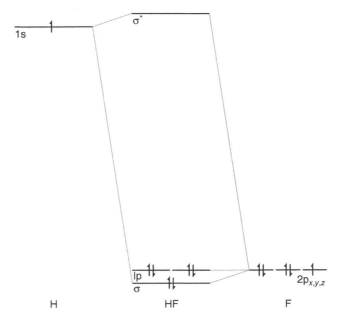

Figure 3.12 MO diagram for the HF molecule.

3.2.7 Polyatomic Molecules

So far we have looked at diatomic molecules, which is a very limited part of chemistry; most molecules have three or more atoms. Many of the conclusions drawn from diatomic molecules can, however, be transferred to polyatomic molecules. For example, the bonding of ethene, C_2H_4, can be described by four CH-σ bonds, one CC-σ bond, and a CC-π bond plus the corresponding antibonding orbitals. Ethene is not linear and the z-component of the orbital angular momentum is not a preserved quantity; so, talking of σ and π bonds is strictly speaking not correct. However, we borrow the nomenclature and consider the *local* symmetry, a σ bond is pointing in the direction of the bond axis and atomic orbitals making a π bond is pointing in a direction perpendicular to the bond axis. In ethene, only one of the 2p components forms a π bond, while the other two forms σ together with the 2s.

A molecular orbital is in general a mixture of atomic orbitals

$$\varphi_i = \sum_k c_{ki} \chi_k, \tag{3.31}$$

where the summation index k runs over all atomic orbitals. The MO coefficients c_{ki} are in general all nonzero except for symmetry reasons, similarly to the separation of σ and π orbitals for diatomic molecules.

Many of the coefficients are, however, very nearly zero; for example, the 1s on carbon does in practice not contribute to any of the bonds in ethene.

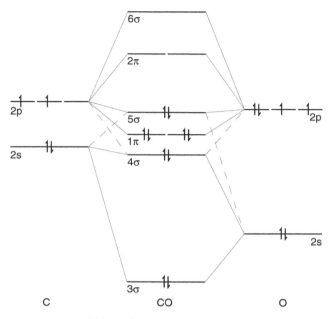

Figure 3.13   MO diagram for the CO molecule.

Atomic orbitals on the same atom as well as on different atoms mix when forming molecular orbitals if there is energy to be gained in the process. There are two criteria to be met for mixing to potentially occur: (i) the orbitals should have similar energy and (ii) the orbitals should be spatially close. The latter criterion does not necessarily imply that all AOs must be close to each other; for example, π systems are often very delocalized, such as in benzene where all six $2p$-π AOs contribute equally to the lowest π MO.

As we go from boron to fluorine in the periodic system, the difference in orbital energy for the 2s and 2p increases, and the amount of 2s-2p mixing in MOs decreases. This property that the 2s and the 2p are close in energy gives carbon its interesting chemistry. The bonding in the CH_4 molecule is conveniently described by four orbitals on carbon that is an equivalent mix of the four valence orbitals, called hybrid orbitals, and in this case sp^3 hybrid orbitals since one 2s and three 2p components are involved. We can write these hybrid orbitals as

$$h_1 = 2s + 2p_x + 2p_y + 2p_z,$$

$$h_2 = 2s + 2p_x - 2p_y - 2p_z,$$

$$h_3 = 2s - 2p_x + 2p_y - 2p_z,$$

$$h_4 = 2s - 2p_x - 2p_y + 2p_z, \tag{3.32}$$

in which case there are four equivalent orbitals pointing to four corners in a cube forming a tetrahedral structure, which is the structure for CH_4, with the 109.47° angle between the bonds. There are two other basic hybrid orbitals that can be formed with 2s and 2p: (i) sp^2 that involves the 2s and two of the three 2p components and (ii) sp that involves the 2s and one of the three 2p components.

Ethene is the archetype for sp^2 hybrid orbitals that form 120° angles between the bonds. The unhybridized 2p orbital on the carbons forms the π bond.

Acetylene is finally the archetype for sp hybrid orbitals with 180° between bonds. Here the two unhybridized 2p orbitals form two π bonds.

3.3 FURTHER READING

For a more detailed description of MO theory, see for example

- Atkins P.W. and DePaula J., "Physical Chemistry", 10th edition, Oxford University Press, Oxford, 2014, ISBN: 9780199697403.
- Atkins P.W. and Friedman R., "Molecular Quantum Mechanics", 5th edition, Oxford University Press, Oxford, 2010, ISBN: 97801199541423.
- Laidler K.J., Meiser J.H. and Sanctuary B.C., "Physical Chemistry", 4th edition, Houghton Mifflin Company, Boston, MA, 2003, ISBN: 0618123415.

4

HARTREE–FOCK THEORY

Hartree–Fock (HF) theory has a special status in the modern quantum chemistry. If someone makes a statistical analysis of the popularity of different theoretical methods in scientific publications over the last decade, with no doubt Hartree–Fock method will not appear at the top. At the same time, it is impossible to find a textbook in quantum chemistry that does not cover HF theory (and this book is not an exception). Based on such discrepancy between the role of the HF theory in educational textbooks and its appearance in a "real life," reader might get a false impression that this theory has only educational and historical value. This is not true.

Hartree–Fock method often gives a very realistic qualitative description of the ground state of a system in an equilibrium geometry. There are few cases where Hartree–Fock theory predicts a completely wrong molecular structure, since it is usually dominated by a single closed-shell electronic structure. Even if Hartree–Fock results are quantitatively different from the experiment, correction schemes can be used to improve the results in a gradual and a systematic way. One can apply perturbation theory (MP2), or approximate treatment of exchange and correlation using density functional theory (DFT), or use Hartree–Fock wave function as a reference function for explicit correlation methods. Most important, the Hartree–Fock method is based on a strict and simple physical model, which makes it a useful and straightforward tool for the theoretical interpretation of the electronic structure and chemical bonding.

Multiconfigurational Quantum Chemistry, First Edition.
Björn O. Roos, Roland Lindh, Per Åke Malmqvist, Valera Veryazov, and Per-Olof Widmark.
© 2016 John Wiley & Sons, Inc. Published 2016 by John Wiley & Sons, Inc.

4.1 THE HARTREE–FOCK THEORY

In this section, we introduce the approximation of the wave function upon which the Hartree–Fock theory is based. Subsequently, we examine the outcome of this approximation for the Schrödinger equation, leading up to the Hartree–Fock equations.

4.1.1 Approximating the Wave Function

The complexity of the Schrödinger equation increases with the number of electrons. In quantum mechanics, the wave function describing N particles depends on positions of all of them simultaneously (or, if the wave function is considered in momentum representation, it depends on all momenta). Because of the complexity of wave function, the Schrödinger equation can be solved exactly only for the simplest model systems, which are of no interest to chemists. For real systems, two kinds of approximations can be applied: simplification of the Hamiltonian (e.g., ignoring some physical interactions), or simplification of the wave function.

Here we explore a wave function model that completely ignores the correlation of the electrons in a system. Such a simplification gives us a wave function for an N-electron system $\Psi(1, 2, \ldots, N)$ in a form of products of independent one-particle wave functions (known as the Hartree product).

$$\Psi(1, 2, \ldots, N) = \psi_1(1)\psi_2(2) \ldots \psi_N(N), \qquad (4.1)$$

where $\psi_k(k) = \varphi_k(k)\sigma_k$ represents the one-particle spin-orbital wave function–the spin-orbitals–for the kth electron with spatial $\varphi_k(k)$ and spin σ_k components.

The solution of the Schrödinger equation for an N-electron system, using the Hartree-product wave function, can be found in a "mean-field" manner. That is, the structure of the Schrödinger equation of the Hartree product is such that each individual electron does not experience the explicit Coulombic repulsion from the other electrons but rather experience a smeared out electrostatic field of all the other electrons–the mean-field. Hence, the electronic energy includes the sum of the kinetic energy of individual electrons, their interaction with the nuclei, and a sum of the interaction of individual electron with the electrostatic field generated by other electrons. The Schrödinger equation for the Hartree product, in a finite basis, can be solved iteratively by optimizing one orbital at the time and continuing until there is no further change of any orbitals–self-consistency.

For a wave function that neglects the electronic correlation, it works surprisingly well for, for example, the ground state of an electron gas. However, it has wrong asymptotic behavior when the particles are close to each other or, in contrary, when the distance between the particles is infinite.

The source of such a critical error of the Hartree-product approximation is the lack of a fundamental symmetry property of the wave function with respect to permutation of two electrons. The electrons are indistinguishable; thus, an exchange of two electrons should not alter the electron density, that is, $|\Psi(\mathbf{r}_1, \mathbf{r}_2)|^2 = |\Psi(\mathbf{r}_2, \mathbf{r}_1)|^2$. There are two possibilities to preserve the density; the wave function can be either symmetric or antisymmetric, with respect to permutation of a pair of electrons. The symmetric wave function is attributed to bosons (particles with integral spin), while antisymmetric wave function is attributed to fermions (particles with half-integral spin). Experimental evidence in the 1920s established that electrons must be fermions and that they obey the so-called Pauli exclusion principle. For a fermionic wave function, one expects the following:

$$\Psi(1, 2, \ldots, n, \ldots, m, \ldots, N) = -\Psi(1, 2, \ldots, m, \ldots, n, \ldots, N). \qquad (4.2)$$

The Hartree product shows no such symmetry with respect to particles permutation. However, one can use a simple mathematical trick, suggested by Fock [1] shortly after the publication of Hartree, and antisymmetrize the wave function by writing it in a determinant form of spin-orbit orbitals as follows:

$$\Psi(1, 2, \ldots, N) = \frac{1}{\sqrt{N!}} \begin{vmatrix} \psi_1(1) & \psi_2(1) & \cdots & \psi_N(1) \\ \psi_1(2) & \psi_2(2) & \cdots & \psi_N(2) \\ \cdots & \cdots & \cdots & \cdots \\ \psi_1(N) & \psi_2(N) & \cdots & \psi_N(N) \end{vmatrix}. \qquad (4.3)$$

The wave function in the form of a determinant (a Slater determinant and sometimes referred to as an electronic structure) is by construction antisymmetric with respect to exchange of two electrons (the exchange of two electrons corresponds to the interchange of two columns in the determinant and results in a change of sign).

The wave function in the form of antisymmetrized product of one-electron orbitals is the basis for Hartree–Fock theory. This approximation is, although having the fundamental antisymmetry property of a fermionic wave function, still a brute simplification as compared to the exact wave function. The main subject of the rest of the book is to explore the qualitative and quantitative limitations of the Hartree–Fock wave function and to develop wave function approximations that correct for this deficiency.

We also note that the Slater-determinant wave function introduces a special type of electron correlation, the spin-correlation. That is, the wave function has zero amplitude if the position coordinates of two electrons with the same spin (alpha or beta) coalesce. This is a manifestation of the Pauli exclusion principle. In Quantum Chemistry text books, the electronic correlation energy, as defined by P.-O. Löwdin, is the difference in energy between the Hartree–Fock wave function and the exact wave function. We have to keep in mind, for future discussions, that this so-called correlation energy does not include contributions due to the spin-correlation already present in the Hartree–Fock wave function.

4.1.2   The Hartree–Fock Equations

In this section, we look in some detail on the Schrödinger equation as it is simplified using a wave function model based on a Slater determinant. The Hamiltonian, describing the electrons in a field of nuclei contains the following terms: (i) Coulomb repulsion between positively charged nuclei, (ii) Coulomb attraction between electrons and nuclei, (iii) the kinetic energy of electrons, and (iv) the Coulomb repulsion between electrons.

The Hamiltonian, in first quantization formalism, for a many-electron system can be written in terms of the zero-, one-, and two-electron terms as

$$\hat{H} = \sum_{I<J} \frac{Z_I Z_J}{r_{IJ}} - \sum_i \frac{1}{2} \nabla_i^2 - \sum_{i,I} \frac{Z_I}{r_{iI}} + \sum_{i<j} \frac{1}{r_{ij}}. \tag{4.4}$$

The first term, the nuclear-repulsion term, which depends only on the internuclear distances, $r_{IJ} = |r_I - r_J|$, and the nuclear charges Z on atoms I and J, contains no electronic coordinates. Hence, we do not need to elaborate on this term any further. The second and third terms, the one-electron terms, depend on the coordinates of the electrons and include two parts: the kinetic energy of the electrons and the attractive electron–nucleus interaction, depending on the electron–nucleus distances $r_{iI} = |r_i - r_I|$. The last and fourth term, the two-electron term, describing the repulsive electron–electron interaction, depends on the pair interelectronic distances, $r_{ij} = |r_i - r_j|$.

Projecting the Schrödinger equation from the right with the normalized wave function, we can express the energy of the wave function as an expectation value of the Hamiltonian.

$$E = \langle \Psi | \hat{H} | \Psi \rangle. \tag{4.5}$$

The one-electron part of the energy, reduces down to

$$E_{(1)} = - \sum_i \left\langle \psi_i \left| \frac{1}{2} \nabla_i^2 + \sum_I \frac{Z_I}{r_{iI}} \right| \psi_i \right\rangle. \tag{4.6}$$

The two-electron part of the energy,

$$E_{(2)} = \left\langle \Psi \left| \sum_{i<j} \frac{1}{r_{ij}} \right| \Psi \right\rangle \tag{4.7}$$

using the expression of n-electron wave function in the form of Slater determinant, reduces down to a summation over pairs of electrons.

$$E_{(2)} = \sum_{i<j} \{ [ij|ij] - [ij|ji] \} = \sum_{i\leq j} \{ [ij|ij] - [ij|ji] \}, \tag{4.8}$$

where $[ab|cd]$ denotes the integral

$$[ab|cd] = \left\langle \psi_a \psi_b \left| \frac{1}{r_{ij}} \right| \psi_c \psi_d \right\rangle = \int \psi_a^*(i) \psi_b^*(j) \frac{1}{r_{ij}} \psi_c(i) \psi_d(j) \, d\tau_i \, d\tau_j,$$

where the integration is made over the spatial and the spin variables.

The first term in the right-hand side of Eq. 4.8 is the classical Coulombic inter-action term, which the Hartree–Fock wave function shares with the Hartree–product wave function, while the second term, the so-called exchange term, is unique to the Hartree–Fock wave function and is a manifestation of the antisymmetry of a fermionic wave function.

We also note that the summation in Eq. 4.8 can be written with an alternative range of the summation. For the additional term–the so-called self-interaction, i and j are equal, and the two integrals are identical, and the term vanishes. We make the following technical note, in approximative methods, based on the right-hand side of Eq. 4.8, in which the pair of integrals are approximated differently and hence do not cancel perfectly, will include a contribution to the total energy that is due to self-interaction. For example, many DFT implementations treat the Coulomb integrals exactly, while the exchange integrals are approximated and are included in the so-called correlation-exchange functionals. Schemes for self-interaction correction (SIC) or the exact removal of the self-interaction via optimized effective potentials (OEP) have been implemented. However, these schemes generate poten-tials which are not uniquely defined, complicated to compute. The improvement in accuracy is not significantly better that for functionals ignoring the self-interaction error and fail to correctly describe some molecular properties.

To proceed and pave the way for a variational optimization of the wave function parameters of the Hartree–Fock wave function (the spin-orbitals), we select to reexpress these two two-electron contributions in terms of two new operators, the Coulomb and the exchange operator. The integrals $[ij|ij]$ correspond to an electrostatic interaction between two electrons. The corresponding operator $\hat{J}_j$, the Coulomb operator, transforms an arbitrary function $f(1)$ as a linear operator:

$$\hat{J}_j f(1) = \left[\int \frac{\psi_j^*(2)\,\psi_j(2)}{r_{12}}\,d\tau_2 \right] f(1). \tag{4.9}$$

While the exchange integrals, $[ij|ji]$, are reexpressed as a nonlocal operator, the exchange operator:

$$\hat{K}_j f(1) = \left[\int \frac{\psi_j^*(2)\,f(2)}{r_{12}}\,d\tau_2 \right] \psi_j(1). \tag{4.10}$$

Let us make it clear here that the two new operators are not equivalent to the Coulomb and exchange integrals but will in a bra-ket formalism reproduce the energy contributions of these two terms. Furthermore, we notice that in the expression for the total Hartree–Fock energy in Eq. 4.8, the exchange integrals have the opposite sign comparing to Coulomb repulsion. It means that the antisymmetry of a fermionic wave function will reduce the repulsion between electrons as compared with a classical model.

The two-electron energy term of the Hartree–Fock energy, Eq. 4.8, can now be rewritten using the Coulomb and exchange operators:

$$E_{(2)} = \frac{1}{2} \sum_{i,j} \langle \psi_i | \hat{J}_j - \hat{K}_j | \psi_i \rangle. \tag{4.11}$$

Furthermore, we now introduce the so-called Fock operator

$$\hat{F} = \hat{h} + \sum_j (\hat{J}_j - \hat{K}_j), \tag{4.12}$$

where $\hat{h}$ is the one-electron operator (kinetic energy and interaction with all nuclei).

Using the variational principle to optimize the Hartree–Fock wave function will produce the so-called Hartree–Fock equations (for details consult, e.g., Refs [2, 3]). Briefly, the minimum of the total energy, with respect to variations of the wave function parameters, corresponds to the condition $\delta E = 0$, subject to the constraint that the spin-orbitals are orthogonal under the same variation (introduced in the Lagrangian expression with the Lagrangian multipliers ϵ_j) that leads us to the equation

$$\langle \delta \psi_i | \hat{F} - \epsilon_i | \psi_i \rangle = 0.$$

This expression should hold for arbitrary variations, $\delta \psi_i$, which leads us to the final set of equations, the Hartree–Fock equations:

$$\hat{F} \, \psi_j = \epsilon_j \, \psi_j. \tag{4.13}$$

Equation 4.13 defines, for the Fock operator ($\hat{F}$), a set of eigenvectors–the canonical Hartree–Fock orbitals, ψ_j, and associated set of eigenvalues, ϵ_j–the Hartree–Fock orbital energies.

At convergence, the HF energy is computed as

$$E = \sum_i \epsilon_i - \frac{1}{2} \sum_{ij} \langle \varphi_i | \hat{J}_j - \hat{K}_j | \varphi_i \rangle. \tag{4.14}$$

The first term includes the sum of all the orbital energies–that is, the kinetic energy, the nuclei–electron attraction energy, and the electron–electron repulsion of a canonical HF orbital according to Eq. 4.12 Since each such term includes all the interactions of a particular electron, let us say i, with all other electrons, there is a double counting of this interaction energy–ϵ_i contains, for example, an energy contribution for the interaction with, let us say, electron j; the same contribution, however, is also included in ϵ_j. The second term in the energy expression eliminates this double counting.

The ability to assign orbital energies to the orbitals is important for simple models and interpretations. Under the assumption that the orbitals do not change under excitation, we would be able to estimate ionization potentials (IP) from the orbital energies, the so-called Koopmans' [4] theorem. In its simplicity, lack of orbital relaxation, and electron correlation, this works remarkably well.

An effect of applying of the Fock operator to an orbital depends on occupancy of the orbital. If an orbital is occupied by an electron, this electron interacts with $N - 1$ remaining electrons. In contrary, an empty orbital has a "hole" that interacts with N electrons. As a result, the energy gap between occupied and virtual orbitals is overestimated in Hartree–Fock theory.

Furthermore, Eq. 4.11 gives us a very simple explanation of why the lowest triplet excited state normally has a lower energy as compared with the lowest singlet excited state. Under the assumption that the orbitals and orbital energies are identical for the two states, the energy difference is due to that there is a difference in spin for the excited electron. Let this electron be denoted i. With the restriction mentioned above, the energy difference between the singlet and triplet state is due to the interaction of this electron with all others:

$$E^i_{(2)} = \sum_j^{2N} \langle \psi_i | \hat{J}_j - \hat{K}_j | \psi_i \rangle.$$

Considering that in the expressions of the Coulomb and the exchange operators there is included an integration over both spatial and spin coordinates, we will have different contributions for the singlet and the triplet state. While for all orbitals that are doubly occupied, we will have the same energy contributions, there is a difference for the interaction with the electron in the orbital, k, from which the excited electron was removed. For the singlet excited state we will have that this energy translates to $[ik|ik]$, while for the triplet state the energy is computed as $[ik|ik] - [ik|ki]$. Hence, the triplet excited state is lower in energy, as compared to the single excited state, due to exchange stabilization. This is the origin of the relative stability difference between the lowest triplet and singlet excited state.

4.2 RESTRICTIONS ON THE HARTREE–FOCK WAVE FUNCTION

While introducing the Hartree–Fock theory we used spin-orbitals, which depends on both spatial and spin coordinates. In this respect, we have so far not put any restrictions on the relationship between the spatial part of the orbitals of electrons formally occupying the same orbital but with different spin. It is possible that during variational solution of the Hartree–Fock equation the spatial parts of the wave function are the same for α and β spin-orbitals. For instance, such solution can be the case if the system is in equilibrium and has an even number of electrons. If no constraint is applied to the shape of the wave function, the HF method is called the Unrestricted HF (UHF) method. However, one can restrict the spatial part of associated spin-orbitals to be identical, which is the basis of Restricted HF (RHF) method, where

$$\varphi^\alpha(\boldsymbol{r}) = \varphi^\beta(\boldsymbol{r}).$$

In the latter case, the Hartree–Fock equations and energy can be reduced to a summation over spatial orbitals, instead of spin-orbitals, for example, the total RHF energy, in the closed-shell case, is expressed as

$$E = 2 \sum_i \epsilon_i - \sum_{ij} \left\langle \varphi_i \left| \hat{J}_j - \frac{1}{2}\hat{K}_j \right| \varphi_i \right\rangle. \qquad (4.15)$$

The Hartree–Fock method is a variational method; thus, the restrictions in the RHF reduce the number of variational parameters. Hence, in general, we will find that the RHF energy is larger than the UHF energy. However, for systems with an even number of electrons, the variational principle normally will lead to the optimal spatial extension of associated α and β spin-orbital is identical. Here there is no difference between the wave functions or the energies obtained by the RHF and UHF approaches. For an odd number of electrons, or for a system with an even number of electrons, but not all in doubly occupied orbitals, the variational procedure applied to the UHF model will render the spatial part of associated alpha and beta electrons to have a different extension in space.

For the RHF method, we have two different cases: (i) the closed-shell and (ii) the (partial) open-shell formalism. The latter is called the Restricted-Open-Shell (ROHF) method, with two variants: high-spin (HS-ROHF) and low-spin (LS-ROHF) referring to if all of the unpaired electrons have the same spin or not. The RHF acronym is today most exclusively used to denote only the first case. Anyhow, in both cases, the constraint is that the spatial part of the associated alpha and beta spin-orbital is identical.

One important property of UHF approach is the ability to break symmetry. One usually assumes that the electronic wave function should have the same symmetry as the symmetry of the nuclear frame. However, it is not always the truth. The electronic wave function can undergo so-called symmetry breaking. Antiferromagnetic ordering is one example of such lowering of the symmetry. Another example is the Jahn–Teller effect, when a distortion of the electronic subsystem is followed by a distortion of the nuclear subsystem. Furthermore, if one considers any molecule with symmetry and remove (or add) one electron, it can lead to a case with a broken symmetry. The unrestricted Hartree–Fock method can easily describe such cases, since the spatial part of the wave function is independent for α and β spin. The only catch here: UHF describes "one half" of the true picture. Changing all α electrons to β and vice versa will give us another but identical solution. A typical case that demonstrates this is the hydrogen molecule at infinite distance; here we have one solution with an α electron at the first hydrogen and a β electron at the other. A permutation of the α and β electrons yields another solution of the UHF equations but with an identical energy. The correct wave function, of course, is a symmetrized version of the two symmetry-broken UHF solutions.

4.2.1 Spin Properties of Hartree–Fock Wave Functions

The spin properties of the wave function are of significant interest. In particular, it is instrumental for the computational chemistry to understand when the solutions generated in a simulation is a spin eigenstate, or if the generated wave function is spin-contaminated. In this section, we look at these aspects in some details for the various Hartree–Fock wave function. In particular, we analyze the property of the HF wave functions with respect to the two spin operators $\hat{S}_z$ and $\hat{S}^2$; the first one is the projection of the spin on the z-axis and the second one is the operator for the magnitude of the spin. Here we remember that if the wave function is a spin eigenfunction, the following two relations hold.

$$\hat{S}^2 \Psi = S(S+1)\Psi,$$

$$\hat{S}_z\Psi = M_s\Psi.$$

First, let us examine the $\hat{S}_z$ operator, the projection of the spin on the Z-axis, and to what extent the RHF, UHF, and ROHF wave functions are eigenfunctions of this operator. The building block of this analysis is based on the fact that the spin-orbitals are eigenfunctions of the single particle spin operator $\hat{s}_z$, that is,

$$\hat{s}_z\psi^\sigma = \pm\frac{1}{2}\psi^\sigma, \tag{4.16}$$

where it takes on the $+$ and the $-$ values for the α and β orbitals, respectively. The n-particle spin operator, $\hat{S}_z$, is expressed as $\hat{S}_z = \sum_i \hat{s}_{zi}$. Since the operator leaves the spatial form of the wave function untouched we note that

$$\hat{S}_z\Psi^{HF} = \frac{(n_\alpha - n_\beta)}{2}\Psi^{HF}, \tag{4.17}$$

where n_α and n_β are the number of electrons with spin α and β, respectively. Hence, the RHF, UHF, and ROHF wave functions are all trivially eigenfunctions to the $\hat{S}_z$ operator.

This leaves us with the more elaborate task, the analysis of $\hat{S}^2$ operator as applied to the different HF wave functions. Here we need some more machinery for our analysis. First, the $\hat{S}^2$ operator, the operator for the total magnitude of the spin, can be expressed, in terms of the $\hat{S}_z$, the raising ($\hat{S}_+$), and the lowering ($\hat{S}_-$) operators, as

$$\hat{S}^2 = \hat{S}_x^2 + \hat{S}_y^2 + \hat{S}_z^2 = \hat{S}_z^2 + \hat{S}_z + \hat{S}_-\hat{S}_+ = \hat{S}_z^2 - \hat{S}_z + \hat{S}_+\hat{S}_-. \tag{4.18}$$

In Eq. 4.18, we used the commutation properties of spin operators, $[\hat{S}_x, \hat{S}_y] = i\hat{S}_z$, and that the raising and lowering operators are expressed as $\hat{S}_\pm = \hat{S}_x \pm i\hat{S}_y$. Typically, the raising and lowering operators change the M_S value of a spin eigenfunction according to $\hat{S}_+|S, M_S> = \sqrt{(S(S+1) - M_S(M_{S+1}))}|S, M_{S+1} >$ (for an electron, this complicated constant reduces down to 1), with the condition that M_S is less than S or larger than $-S$ for the raising and the lowering operator, respectively; otherwise, the result is that the wave function is annihilated. When we continue here, we will look at what the action of $\hat{S}^2$ is on the wave function. For the wave function to be an eigenfunction, the wave function should be invariant times an arbitrary constant.

Since any HF wave function is an eigenfunction of $\hat{S}_z$, we note that the first two terms in Eq. 4.18 are unproblematic–these operators will leave the HF wave function unchanged. We have to examine the result of the last operator in Eq. 4.18 to see what HF wave functions are spin eigenfunctions.

Expressed in terms of the operators of the individual electrons,

$$\hat{S}_-\hat{S}_+ = \sum_i \hat{s}_-(i) \sum_j \hat{s}_+(j). \tag{4.19}$$

Let us split the sum into two parts:

$$\sum_{ij} \hat{s}_-(i)\hat{s}_+(j) = \sum_i \hat{s}_-(i)\hat{s}_+(i) + \sum_{i \neq j} \hat{s}_-(i)\hat{s}_+(j). \tag{4.20}$$

The $\hat{s}_+(i)$ operator applied to any ψ_i^α spin-orbital gives 0, since there is no higher value of m_s for an α electron. While if the $\hat{s}_-(i)$ operator is applied to a ψ_i^α spin-orbital, it initially yields

$$\hat{s}_-\psi_i^\alpha = \sum_j \left\langle \varphi_i^\alpha | \varphi_j^\beta \right\rangle \psi_j^\beta, \tag{4.21}$$

where $\langle \varphi_i^\alpha | \varphi_j^\beta \rangle$ is the overlap matrix between the spatial parts of the spin-orbitals. Now we have to remember that the orbitals the operators act on are orbitals in a determinant expression–the operator actually acts on the determinant–and that the Pauli exclusion principle holds. That is, the summation in Eq. 4.21 is limited to the unoccupied β orbitals–we cannot put a β electron in an already occupied orbital; if we do so, the determinant has the value zero. We further note that in the restricted case (same spatial part for associated α and β electrons) Eq. 4.21 reduces to $\hat{s}_-\psi_i^\alpha = \psi_j^\beta$.

The effect of the $\hat{s}_-(i)$ operator is the opposite; a ψ_i^β spin-orbital becomes equal to 0, and the effect of $\hat{s}_+(i)$ on a ψ_i^β spin-orbital changes the spin to α and results in a summation, as in Eq. 4.21, now restricted to unoccupied α spin-orbitals.

Hence, for the first term in Eq. 4.20, we will have that the operator first flips all β/α electrons to α/β, with the restriction that the spatial part is not already occupied, and then flips them back. This leaves the wave function unchanged, except for a multiplicative factor, or the operator will make the wave function vanish. For the second term, the first flip will be followed by a second flip that does not undo the action of the first operator.

Let us look at the implications this has for the RHF, the high-spin (HS), and low-spin (LS) versions of the ROHF, and the UHF wave functions, respectively.

First, since the RHF wave function has all orbitals doubly occupied $\hat{S}_\pm$ acting on the wave function will make both terms in Eq. 4.20 vanish, that is the RHF equation is a spin eigenfunction. Second, for the HS-ROHF wave function, let us say that all unpaired electrons have α spin, we find that the operators in Eq. 4.20 will make the contribution vanish. Thus trivially for the doubly occupied space and for the unpaired orbitals, we start by applying the raising operator. That is, the HS-ROHF wave function is a spin eigenfunction too. Third, for the LS-ROHF, where we have both unpaired α and β electrons, we have that operations on the doubly occupied orbitals vanish. Let us say we start by using the raising operator we will have that operating on the unpaired α electrons will make the wave function vanish, while for the unpaired β electrons these will now be spin-flipped to an unpaired α electrons. The application of the second operator will for the first term in Eq. 4.20 undo the flip, while for the second term the operator will flip another unpaired α electron into a new unpaired β electron. That is, the LS-ROHF wave function is not invariant with respect to the second operator in Eq. 4.20 and is thus found to not be a spin eigenfunction. Fourth and final, for the UHF wave function, the result of applying the raising operator will trivially vanish as we operate on the α electrons. For the β electrons, the raising operator will spin-flip them into the α orbital space that is unoccupied. However, as for the LS-ROHF, the subsequent lowering operator in the second term of Eq. 4.20 will not undo this change. Hence, the UHF wave function is not a spin eigenfunction.

The deviation of the expectation value of $\langle S^2 \rangle$ from the correct value, $(S(S+1))$, (the so-called spin contamination) can in the case of a UHF wave function be computed by projecting Eq. 4.20 from the left with the same wave function, giving

$$\langle S^2 \rangle - \langle S^2 \rangle^{\text{correct}} = \sum_i^{\min(N_\alpha, N_\beta)} 1 - \sum_{ij}^{N_\alpha, N_\beta} |\langle \varphi_i^\alpha | \varphi_j^\beta \rangle|^2$$

$$= N_\beta - \sum_{ij} |\langle \varphi_i^\alpha | \varphi_j^\beta \rangle|^2$$

assuming that number of electrons with spin β does not exceed number of electrons with spin α. If the spatial parts of the wave function are identical, as it is the case of RHF and ROHF, the overlap matrix is unity matrix, and spin contamination vanishes. In the case of a true UHF wave function, this expression is always different from 0, and thus the wave function is not an eigenfunction of the $\hat{S}^2$ operator. For the cases with even number of electrons and molecule at an equilibrium distance, the difference between the spatial parts of the wave functions is small, and the spin contamination (the difference between exact and computed values for $\langle S^2 \rangle$) is negligible. In other cases, the spin-contamination for UHF wave function cannot be ignored; however, for doublets and triplets it tends to be a good approximation.

For further details, we refer the reader to Refs [2, 5]. We note that the UHF wave function is used regularly. This can be done safely if the user checks the value of the expectation value of $S(S+1)$ and monitors that it does not deviate too much as compared to the expected spin eigenvalue of the state.

4.3 THE ROOTHAAN–HALL EQUATIONS

Hartree–Fock equation is a significantly simplified version of the Schrödinger equation; however, it can only be solved analytically for some of the most trivial cases. In general case, however, a numerical solution of HF equations can be obtained if a basis set is used to describe the spatial orbitals. The Linear Combination of

Atomic Orbitals (LCAO) expansion has been introduced in Chapter 3, so that the spatial part of a molecular spin-orbital, $\varphi_i(r)$, can be presented as a linear combination of atomic orbitals (or Gaussian basis functions) $\chi_p(r)$:

$$\varphi_i(r) = \sum_p C_{pi}\chi_p(r). \tag{4.22}$$

The particular shape of the basis functions is described in Chapter 6.

Using the LCAO approach, we transform the Hartree–Fock equations to the language of linear algebra. Instead of solving the complicated integro-differential HF equations, one rather minimizes the energy with respect to the N unknown coefficients C_{ip}. The size of the problem now depends on the size of the basis set: the larger the basis set the larger the number of unknown coefficients.

By right and left projections with the basis set, χ_p, we can now reformulate the HF equations as follows:

$$\mathbf{FC} = \mathbf{SCE}. \tag{4.23}$$

These equations are the so-called Roothaan–Hall equations [6, 7], where the elements of the Fock matrix are computed as

$$F_{pq} = \langle \chi_p | \hat{F} | \chi_q \rangle$$

and the corresponding elements of the overlap matrix are

$$S_{pq} = \langle \chi_p | \chi_q \rangle.$$

The solution of the Roothaan–Hall equations are subject to the orthonormality constraint

$$\mathbf{C}^\dagger \mathbf{SC} = 1$$

and on convergence we find that

$$\sum_{p,q} C_{pa}^* F_{pq} C_{qi} = (\mathbf{C}^\dagger \mathbf{FC})_{ai} = 0 \tag{4.24}$$

that is, there is no interaction between the occupied and virtual orbitals.

The solution to the secular equation provides us with sets of coefficients that equals in size to the number of basis function. However, we need only those solutions that corresponds to the occupied orbitals. These additional orbitals, the so-called virtual orbitals, do not enter the energy expression and should be used for other purposes with caution. Furthermore, we again remind that the Fock matrix implicitly depends on the occupied orbitals. Hence, the solution of the secular equation has to be iterated until the very same orbitals, which are used in the construction of the Fock matrix, are found to be the solution to the secular equation. Typically, we denote the solution of the Roothaan–Hall equations as the Self-Consistent Field (SCF) solution, a solution to the HF equations in a *finite* basis set, while the Hartree–Fock solution is reserved to denote the SCF solution in an **infinite** basis.

The self-consistent procedure for the solution of the Hartree–Fock equations in the matrix form includes the following steps: (i) a trial wave function is used to construct an initial set of occupied orbitals, (ii) these orbitals are used to construct the Fock matrix, and (iii) the solution of Roothaan–Hall equations gives a new set of orbitals. The procedure is repeated from step (ii) until the difference between the orbitals used in step (ii) and those generated in step (iii) are smaller than a given threshold. This procedure can be improved by using various algorithms that either accelerate the convergence, or stabilize the solution (for details, consult Refs [8–11]). There is, however, one important issue in these recipes that are used for solving Hartree–Fock equations. The selection of the initially occupied orbitals that are used to construct the starting Fock matrix. These orbitals define implicitly an electronic configuration and require a delicate selection to lead to convergence to the correct state.

4.4   PRACTICAL ISSUES

In the following, we discuss some practical issues with respect to the Hartree–Fock wave functions. First, we discuss the qualitative failure of HF theory to correctly describe bond dissociation. Second, we discuss the existence of multiple solutions to the Roothaan–Hall equations and some common problems of the HF model with respect to symmetry and degeneracy.

4.4.1   Dissociation of Hydrogen Molecule

The equilibrium geometry of molecules is described by Hartree–Fock theory surprisingly well. If we ignore quantitative discrepancies, it would be hard to find a case, where HF predicts a completely wrong geometry for the ground state of a molecule. There are a few cases where HF predicts repulsive interaction for weakly bounded molecules, such as Be_2. But it is fair to say that these molecules are exotic.

The explanation for this success of the HF wave function to model the optimal geometry of an 'ordinary' molecule is the fact that at these structures, the wave function usually is dominated by a single electronic configuration. However, although the energy retrieved by the HF method is rather accurate, usually correct within 1% of the exact energy, it is not accurate enough to get the relative energy of chemical reactions correct. Furthermore, applying the Hartree–Fock model to nonequilibrium systems leads to qualitatively wrong results. This qualitative problem of the Hartree–Fock theory is exemplified by the dissociation of the H_2 molecule (as mentioned in Chapter 1). The total energy of the two hydrogen atoms at infinite interatomic distance, which should be twice that of a single hydrogen atom (-0.5 au), computes to -0.7 and -1.0 au at the RHF and the UHF levels of theory, respectively! This error is both quantitative and qualitative for the RHF method, while only qualitative for the UHF method. Let's take a detailed look at the difference between the description of two hydrogen atoms on short and long interatomic distance. Remember, the construction of the Fock matrix depends on the occupied orbitals. While we in the RHF force the spatial part of the orbital occupied

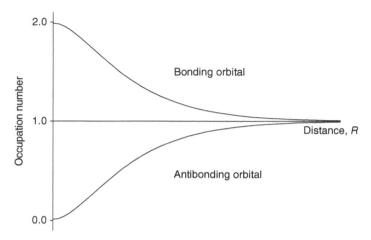

Figure 4.1   Occupation numbers for a pair of bonding and antibonding orbitals.

by the α and β electrons to be identical, we do not have this constraint for the UHF. For the RHF solution, we have the occupied orbital, which is the bonding orbital–a delocalized orbital, while for the UHF method, we have optionally localized atomic orbitals. The former is wrong due to the static inclusion of ionic terms in the total energy, due to the fact that occupied orbitals are delocalized and doubly occupied for all interatomic distances. A more realistic dependence of the occupation numbers for the bonding–antibonding orbital pair is shown in Figure 4.1. As the two hydrogen atoms separate the bonding and antibonding orbitals get degenerate and at infinite distance we should expect equal occupation. This flexibility would allow for the elimination of ionic electron configurations at longer bond distances.

In order to provide for such a description of the H − H bonding, one should allow for several electronic configurations at the same time. From the formal point of view, unrestricted Hartree–Fock wave function performs much better for the dissociation of H_2 molecule. It predicts the correct energy for two noninteractive hydrogen atoms. However, the dissociation corresponds to a solution with a broken symmetry (the UHF solution is not a spin eigenfunction, see 4.2.1 for details): one atom contains an electron with spin α and another–with spin β. Obviously, there is another identical solution, where the first hydrogen atom has a β electron, and the second one has an α electron. A proper description of the noninteractive hydrogen atoms should include a linear combination of these two electronic configurations–a multiconfigurational wave function.

4.4.2   The Hartree-Fock Solution

The solution of the Hartree–Fock equations is found by an iterative procedure that corresponds to a minimization of the energy with respect to variations of the wave function parameters, the orbitals. This solution, however, is not unique. Depending on the initial starting orbitals, one can converge the Hartree–Fock equations to a multitude of different stable solutions. For example, if the initial starting orbitals correspond to an electronic configuration of an excited state, it is possible that the obtained self-consistent solution will correspond to this excited state. There are several techniques that might be useful in searching for a global minimum for the energy in Hartree–Fock equations, but it is fair to say that none of these techniques can guarantee that one obtains a solution that corresponds to the ground state.

If we don't know the ground state electronic configuration, we can artificially partially populate all orbitals, in order to effectively remove the bias toward a particular starting electronic configuration, One such recipe is the so-called temperature Aufbau method. In this method, instead of using fixed integer occupation numbers, we associate partial occupation numbers to both the occupied and virtual orbitals. These occupation numbers depend on the associated orbital energy and a temperature. In the iterative procedure, we gradually reduce the temperature down to zero and reestablish fixed integer occupation numbers consistent with HF theory. In most cases, this iterative procedure will find the ground state electronic configuration. In case of degeneracy, it will not converge, helping us to understand which different electronic configurations are degenerate.

Moreover, the HF method has problems with spatially degenerate orbitals. For example, consider case of the benzene cation; here we remove an electron from the pair of the two degenerate doubly occupied e_{1g} orbitals. In the UHF, we will arbitrarily select one. The resulting Fock operator will now not treat these orbitals in an equal way and ultimately the UHF treatment will result in a symmetry-broken solution and a molecular structure not confined to D_{6h}. By removing a second electron, we can get the RHF model to do the same.

In practice, everyone who applies Hartree–Fock technique to real systems knows that sometimes the convergence in SCF procedure is fast and straightforward; but in other cases, it is either too slow or even diverging. There are various techniques for acceleration or stabilization of self-consistent convergence. We refer our reader to [8–11]. However, it is important to understand that Hartree–Fock theory tries to converge the wave function of a single electronic configuration, which might not be possible at all or be qualitatively wrong. Hence, in case of poor convergence or none, we should take this as an indication of that the HF model is not well suited for our system under investigation. Understanding the reason will allow us to select a more appropriate wave function model.

4.5 FURTHER READING

For more information about Hartree–Fock theory we can recommend the following textbooks:

- J. Almlöf and R. Ahlrichs, Notes on Hartree–Fock theory and related topics, in "European Summerschool in Quantum Chemistry, Book I", ed. P.-O. Widmark, Lund University, Lund, 2011.

- T. Helgaker, P. Jørgensen, and J. Olsen, "Molecular Electronic-Structure Theory", John Wiley & Sons, Ltd, Chichester, 2000, ISBN: 0-471-96755-6.
- R. McWeeny, "Methods of Molecular quantum mechanics", 2nd edition, Academic Press, London, 1992, ISBN: 0-12-486551-8.

4.6 REFERENCES

[1] Fock V. Näherungsmethode zur Lösung des quantenmechanischen Mehrkörperproblems. Z Phys 1930;61:126–148.

[2] Szabo A, Ostlund NS. Modern Quantum Chemistry. Introduction to Advanced Electronic Structure Theory. Mineola (NY): Dover Publications, Inc.; 1996.

[3] Mayer I. Simple Theorems, Proofs, and Derivations in Quantum Chemistry (Mathematical and Computational Chemistry). New York: Kluwer Academic/Plenum Publishers; 2003.

[4] Koopmans T. Über die Zuordnung von Wellenfunktionen und Eigenwerten zu den Einzelnen Elektronen Eines Atoms. Physica 1934;1:104–113.

[5] McWeeny R. Methods of Molecular Quantum Mechanics. London: Academic Press; 1992.

[6] Roothaan CCJ. New developments in molecular orbital theory. Rev Mod Phys 1951;23:69–89.

[7] Hall GG. The molecular orbital theory of chemical valency 8. A method of calculating ionization potentials. Proc R Soc London, Ser A 1951;205:541–584.

[8] Høst S, Olsen J, Jansík B, Thøgersen L, Jørgensen P, Helgaker T. The augmented Roothaan-Hall method for optimizing Hartree-Fock and Kohn-Sham density matrices. J Chem Phys 2008;129:124106.

[9] Rohwedder T, Schneider R. An analysis for the DIIS acceleration method used in quantum chemistry calculations. J Math Chem 2011;49:1889–1914.

[10] Levitt A. Convergence of gradient-based algorithms for the Hartree-Fock equations. ESAIM: Math Modell Numer Anal 2012;46:1321–1336.

[11] Garza AJ, Scuseria GE. Comparison of self-consistent field convergence acceleration techniques. J Chem Phys 2012;137:Art. 054110.

5

RELATIVISTIC EFFECTS

5.1 RELATIVISTIC EFFECTS ON CHEMISTRY

Chemistry is based on empirical rules, and so relativistic effects cannot be defined except as discrepancies compared to predictions of nonrelativistic models. To be meaningful, these models must also be assumed to be accurate enough in all respects except for the lack of relativity. It is then not surprising that consideration of relativistic effects has not become important in chemistry until the use of such accurate models became widespread, especially since open shells and a multitude of low-lying states often make a multiconfigurational picture mandatory. High-energy X-ray spectroscopy that probes the inner shells was easier to treat but with too poor resolution for chemical relevance.

Nevertheless, there are some purely empirical observations of trends across the periodic table, where simple rules seem to hold for lighter elements but not for the heavier ones, which can be easily explained as effects of relativity. Traditionally, these include, for example, the tendency for the $6s^2$ electron pair to remain as atomic shells rather than to partake in bonding in some heavy atom complexes, trends in atomic radii, etc. For the elements, some nonnumerical observations, for example, the color of gold, and the fact that mercury is a liquid, are similarly explained by relativity changes of 6s orbital energy and extent. For a review of such effects, see, for example, Ref. [1].

Multiconfigurational Quantum Chemistry, First Edition.
Björn O. Roos, Roland Lindh, Per Åke Malmqvist, Valera Veryazov, and Per-Olof Widmark.
© 2016 John Wiley & Sons, Inc. Published 2016 by John Wiley & Sons, Inc.

The largest effects can be seen in spectroscopy, where also the earliest atom models were sufficient to show the effects of neglecting relativity. The largest such effect is from changed balance between the kinetic energy of the electrons, which increases for the more compact atomic orbitals, and their potential energy that grows rapidly when the electron gets close to the nucleus. The conventional nonrelativistic expressions for the kinetic and potential energies of a single electron in the Coulomb field of a nucleus are, respectively, $T_{NR} = \mathbf{p}^2/(2m)$ and $V = -Z/r$, while relativistically, the kinetic energy is given by

$$T = \frac{2}{1 + \sqrt{1 + 2T_{NR}/(mc^2)}} T_{NR}.$$

Relativistically correct quantum mechanics is more complicated than this, but it is correct that the relativistic kinetic energy is reduced by a factor that becomes significantly smaller than 1 when the nonrelativistic kinetic energy approaches or exceeds the rest energy of the electron, mc^2, which is on the order of 0.5 MeV or 12 000 000 kcal/mol. This large number has mislead many to assume that the effect on chemistry is insignificant, but this reduction of the kinetic energy has important consequences: the electron distribution contracts near the nucleus, increasing the screening of nuclear charge as seen by the outer electrons. The valence s orbitals contract, due to their penetration to the core region, and their orbital energy decreases, while d orbitals get destabilized by the increased screening. For mercury ($Z = 80$), the extent $\langle r \rangle$ of the 6s orbitals is reduced by 15% compared to its nonrelativistic value. At the same time, the 5d orbitals increase a few percent. The 6s orbital maximum is still outside that of 5d, but the repulsion from the inert 5d shell makes it less easy for another atom to come close enough to form a bond with the 6s orbital.

According to the extensive review by Pyykkö [1] on relativistic effects on chemistry, the relativistic contraction of s and p shells, the subsequent expansion of d and f from screening, and the well-known spin–orbit (SO) splitting are all three effects of comparable magnitude.

Another effect is seen in the relativistic shortening of many bonds. This is however described by Pyykkö [1] as an effect parallel to, but not caused by, the contraction of orbitals. Thermochemistry is affected when a reaction can change the s occupation, and the reaction $PbH_4 \rightarrow PbH_2 + H_2$ is predicted to be endothermic by nonrelativistic calculation, but is correctly described as exothermic once relativistic effects are included [2].

The lowering of the kinetic energy of the electrons is easily included in conventional quantum-chemistry programs and is often called *scalar* relativity. Apart from energetics, the breaking of spin symmetry allows reactions to proceed over barriers between states of different spin multiplicity. Even reactions that do not directly affect bonds to a heavy atom can be promoted by the presence of a heavy atom in a molecule.

The review article by Pyykkö [1] is still a good source for relativistic effects on chemistry and gives further references, for example, to such phenomena as rate-determining intersystem reactions, that is, those which get reaction probability by the spin-symmetry breaking of spin–orbit interactions, also for lighter elements and organic chemistry. Large effects of spin–orbit splitting can occur with dissociation to radical products. Even lighter atoms have a significant energy difference between, for example, $P_{1/2}$ and $P_{3/2}$ states, affecting dissociation energies as well as the distant part of the potential curves (e.g., Cl: 2.5 kcal/mol, Br: 10.5 kcal/mol).

The nature of the bonds can change quite drastically with the heaviest atoms. For instance, two atomic p_σ orbitals may form a σ bond, but not only does scalar relativity change the energetics by affecting the possibility of sp hybridization, but also the SO interaction breaks the degeneracy of the p orbitals. The $p_{1/2}$ and $p_{3/2}$ atomic orbitals have different energy, different ability to hybridize, and different directionality. However, the SO interaction is too weak to overcome the electrostatic quenching of the molecular field for all but the very heaviest atoms, such as U. The effect still shows up, for example, as the weakening of the TlH bond as compared to nonrelativistic expectations, while for U the $7p_{1/2}$ is chemically inert, but the $7p_{3/2}$ is valence [3].

Basically, the spin–orbit interaction is the coupling term in the Hamiltonian between the magnetic moment of the electrons and the magnetic field caused by their motion. Perturbationally, it is typically derived from the Breit–Pauli Hamiltonian, which for an atom is

$$\hat{H}^{SO} = \frac{1}{2c^2} \left(\sum_i \frac{Z}{r_i^3} \hat{\mathbf{l}}_i \cdot \hat{\mathbf{s}}_i - \sum_{ij} \frac{1}{r_{ij}^3} \hat{\mathbf{l}}_{ij} \cdot (\hat{\mathbf{s}}_i + 2\hat{\mathbf{s}}_j) \right),$$

where the angular momentum operators are $\hat{\mathbf{l}}_i = \hat{\mathbf{r}}_i \times \hat{\mathbf{p}}_i$, $\hat{\mathbf{l}}_{ij} = (\hat{\mathbf{r}}_i - \hat{\mathbf{r}}_j) \times \hat{\mathbf{p}}_i$, and the spin $\hat{\mathbf{s}}_i$, and $\hat{\mathbf{r}}_i, \hat{\mathbf{p}}_i$ are the position and momentum operators of the electrons. Atomic units are used, and then $c = 1/\alpha \approx 137.036$.

This is the lowest (c^{-2}) term of a divergent perturbation series. The one-electron term is the spin-same-orbit and the second is the two-electron spin-other-orbits term. There is also a two-electron spin–spin term that is of lesser importance. Of course, these terms originate from relativistic Quantum Mechanics, since they vanish when $1/c \rightarrow 0$. The spin-same orbit term is easily obtained from a central-field solution to the Dirac Hamiltonian, while the two-electron terms similarly come from the Breit Hamiltonian. Very similar expressions are obtained in modern variational two-component Hamiltonians, and then also the Z/r_i^3 factor is replaced by the gradient of the mean-field potential from nuclei and electrons.

The spin of the electrons is always $s = 1/2$, but the vector quantity $\hat{\mathbf{s}}$ cannot be treated as a constant. Scalar products such as, for example, $\hat{\mathbf{s}} \cdot \hat{\mathbf{l}}$ may be evaluated by a "polarization formula":

$$\hat{\mathbf{s}} \cdot \hat{\mathbf{l}} = \frac{\hat{\mathbf{j}}^2 - \hat{\mathbf{l}}^2 - \hat{\mathbf{s}}^2}{2}$$

by defining $\hat{\mathbf{j}} = \hat{\mathbf{l}} + \hat{\mathbf{s}}$.

5.2 RELATIVISTIC QUANTUM CHEMISTRY

The spin–orbit interaction is also directly obtained as part of a relativistic calculation. The field of computational relativistic quantum mechanics had, until perhaps three or four decades ago, been dominated by fairly complex issues that prevented application of its methods to all but the very smallest molecules. The one-electron problem of mean-field approaches such as Dirac–Fock is similar to Hartree–Fock but requires the use of four-component orbitals.

The (time-dependent) Schrödinger equation for an electron is

$$i\frac{\partial}{\partial t}\Psi = \hat{H}\Psi = \left(-\frac{1}{2}\nabla^2 + V\right)\Psi$$

where $V = -e\Phi$, where $e = 1$ and Φ is the electrostatic potential field, using atomic units. The Dirac equation is formally similar, but the wave function has four components, and the Hamiltonian is a 4×4 matrix:

$$i\frac{\partial}{\partial t}\begin{pmatrix}\Psi_1\\\Psi_2\\\Psi_3\\\Psi_4\end{pmatrix} = \begin{pmatrix}mc^2 - e\Phi & 0 & c\hat{\pi}_z & c(\hat{\pi}_x - i\hat{\pi}_y)\\0 & mc^2 - e\Phi & c(\hat{\pi}_x + i\hat{\pi}_y) & \hat{\pi}_z\\c\hat{\pi}_z & c(\hat{\pi}_x - i\hat{\pi}_y) & -mc^2 - e\Phi & 0\\c(\hat{\pi}_x + i\hat{\pi}_y) & \hat{\pi}_z & 0 & -mc^2 - e\Phi\end{pmatrix}\begin{pmatrix}\Psi_1\\\Psi_2\\\Psi_3\\\Psi_4\end{pmatrix},$$

where mc^2 is the rest mass energy ($c \approx 137.036$ in au), $-e\Phi$ is the same V as in Schrödinger's equation, and $\hat{\pi} = \hat{p} + eA = -i\nabla + eA$ can be called the mechanical momentum. A is the electromagnetic vector potential, and this is the reason that we write here $-e\Phi$ instead of V, so the scalar and vector potential appears together in the same formula.

Of course, the equation is not very transparent as it stands, and it is conveniently blocked up as two coupled equations, each for a two-component spinor wave function. The upper one is then Ψ_L, the "larger" wave function, and its two components correspond to the familiar α and β spins. The "smaller" Ψ_S wave function has a small value for solutions corresponding to ordinary electrons with moderately large kinetic energy. The equations are then

$$i\frac{\partial}{\partial t}\begin{pmatrix}\Psi_L\\\Psi_S\end{pmatrix} = \begin{pmatrix}mc^2 - e\Phi & c(\sigma \cdot \hat{\pi})\\c(\sigma \cdot \hat{\pi}) & -mc^2 - e\Phi\end{pmatrix}\begin{pmatrix}\Psi_L\\\Psi_S\end{pmatrix},$$

where the vector of Pauli matrices $\sigma = (\sigma_x, \sigma_y, \sigma_z)$ is used; its components are Cartesian components, the scalar product with the vector operator $\hat{\pi}$ gives a scalar operator that is a single 2×2 matrix, which acts on Ψ_L or Ψ_S. For short, it is written as $\hat{H} = \beta mc^2 - e\Phi + c\alpha \cdot \hat{\pi}$, where β is a 2×2 matrix whose elements are 2×2 matrices, and $\hat{\alpha}$ is a vector with three Cartesian components, of which each is again a matrix of matrices:

$$\beta = \begin{pmatrix}I_2 & 0\\0 & -I_2\end{pmatrix}, \quad \alpha = \begin{pmatrix}0 & \sigma\\\sigma & 0\end{pmatrix}.$$

For a one-electron atom, it is virtually exact, and its hydrogenic solutions are well known. For a point nucleus, it gives then the familiar picture with almost the same states as a nonrelativistic equation, but with the exception of s-states, they are split up by spin–orbit interaction. $j = l + s$ is the total electronic spin and is conserved for all central potentials. For the hydrogenic case, the p-states, and higher, are split up into $p_{1/2}$ and $p_{3/2}$, etc. The magnitude of model errors is shown by the Lamb shift, that is, the $2s_{1/2}$ and $2p_{1/2}$ come out as degenerate by Dirac's equation, but are experimentally seen to be distinct, with a splitting on the order of 10^{-4} kcal/mol. The Dirac equation itself is good enough for chemical applications.

The Dirac–Fock equations are now obtained once a mean-field approximation to the electron–electron repulsion has been formulated. It is often possible to use simply the nonrelativistic Coulomb interaction. A more accurate form adds the Breit interaction ("Dirac–Fock–Breit").

However, modern Quantum Chemistry prefers anyway not to work directly with partial differential equations. Using variational basis set representations, other difficulties appear, but have been essentially solved, for the purpose of chemistry at least. The problem was that when a finite basis set is applied variationally to the Dirac equation, any error in the size of the small component can result in negative energy contributions, getting amplified by an energy minimization procedure. This will result in orbitals with lower energy that should be possible for any of the true positive-energy eigenstates, causing spurious "ghost states" that do not correspond to any physical states. This problem can be solved by making sure that the basis set for the negative-energy states is devised according to a so-called "kinetic balance" criterion.

The comparably simple electrostatic repulsion interaction was now replaced by several tensorial operators, and while single-atom calculations had been performed for a long time with great success, a look into text books or articles seemed to require great insight into angular coupling schemes. The energy minimization used to define orbitals and to devise basis sets appeared to be on somewhat shaky grounds, since the Dirac–Fock operator allows a continuum of negative-energy eigenvalues, which can give problems with variational collapse. The electron–electron interaction contained singular terms, giving a Hamiltonian that was not variationally stable; even discarding these, and using just the nonrelativistic Coulomb interaction, still gave the possibility of variational collapse through the so-called "Brown–Ravenhall disease" (which was however not a problem for SCF-type calculations). Sucher [4] showed how this problem could be solved by using projectors for the proper positive-energy states.

By this time, many issues had been resolved, at least with regard to applications to chemistry. Methods had been found that effectively transformed the relativistic to a new set of two-component equations, much like the conventional nonrelativistic ones, and where just a conventional basis set suffices. For energetically acceptable accuracy, the interaction between electrons could be replaced with the usual electrostatic interaction. The spin–orbit interaction could be separated from the scalar relativity part of the problem, resulting in a one-component formulation that is identical to the conventional, nonrelativistic one, except that the integrals over the basis functions are modified. All the conventional methods of quantum chemistry could, and were, adapted to calculations using either two-component or scalar form, and it also turned out that, in most cases, the spin–orbit interaction could be added in afterward.

The transformations to two-component form is usually done by some form of regular perturbation theory or by the Douglas–Kroll–Hess (DKH) transformation [5]–[7]. For the DKH Hamiltonian, see below.

The term "regular" perturbation theory is generally used to denote perturbation expansion of nondegenerate eigenstates, and its opposite, "singular" perturbation theory, is then degenerate perturbation theory. In the specific case of relativistic quantum chemistry, it is mostly used for a perturbative decoupling, producing a two-component Hamiltonian, which has as the starting point the "CPD Hamiltonian" [8]. These expansions were developed by van Lenthe and coworkers, who used the term "regular approximation," and the starting point is then called the ZORA, or "zeroth-order regular approximation," while the next is FORA (first-order regular approximation) [2]. Note that this approach is relativistic already at zeroth order. Another approach by Rutkowski [9] and Kutzelnigg [10] has been termed Direct Perturbation Theory.

The various methods for including relativistic effects have different advantages and disadvantages; so, the preferred method to implement depends on the type of quantum chemistry program and on the application. A very useful combination is DFT together with the ZORA Hamiltonian, since in DFT the exchange-correlation potential typically is computed on a grid and used in numerical integration. Since the ZORA modification of the Hamiltonian is also very simple to compute as numerical values on a grid, rather than using analytic formulas for matrix elements over the basis functions, DFT and ZORA fit together as hand in glove.

Finally, an important tool is the relativistic effective core potentials, where only the valence electrons are treated by the quantum chemistry program. These can also include core correlation, core polarization, and valence relativistic effects.

Multiconfigurational calculations can be done using the same programs as for nonrelativistic calculations, for example, RASSCF, CASPT2, and RASSI. The most common inclusion of relativity is then through the Douglas–Kroll–Hess method, and we will now specialize to that class of programs.

5.3   THE DOUGLAS–KROLL–HESS TRANSFORMATION

The Douglas–Kroll (DK) transformation is a sequence of unitary transformations that remove the coupling of the large and small components of the Dirac one-electron Hamiltonian through some order in the one-electron external potential $\hat{V}$.

An ∞-order DK transformation achieves an exact splitting of the Dirac Hamiltonian into two uncoupled two-component parts, one for the positive-energy and one for the negative-energy orbitals. Only the positive-energy orbitals are used in the calculation. This can be achieved by an iterative scheme, or it can be truncated at second or third level of accuracy; see also Chapter 6.

The lowest-order Hamiltonian is obtained by the transformation to a representation where energy, momentum, and helicity are simultaneously diagonalized for the free fields. The external potential, after transforming, yields the relativistic corrections.

The Dirac Hamiltonian is customarily written as

$$\hat{H} = \hat{V} + E_0\hat{\beta} + c\hat{\mathbf{p}}\hat{\boldsymbol{\alpha}}, \tag{5.1}$$

where $\hat{V}$ is the external potential, $E_0 = m_e c^2$ is the rest mass energy of the electron, $\hat{\mathbf{p}}$ is the momentum operator, while $\hat{\beta}$ and $\hat{\boldsymbol{\alpha}}$ are standard 4×4 matrices used in relativistic quantum mechanics; see, for example, Ref. [11] or any other textbook on this subject.

The free-field Hamiltonian is the same without the term $\hat{V}$. It is diagonalized by the unitary transformation matrix

$$\hat{U} = (2E_p(E_0 + E_p))^{-1/2} (E_0 + E_p + c\hat{\mathbf{p}}\hat{\boldsymbol{\alpha}}\hat{\beta}) \tag{5.2}$$

in a basis of plane helicity waves.

Hess [6, 7] suggested that a suitable basis set would allow a matrix representation of the operators $\hat{A}$ and $\hat{R}$, which are algebraic functions of $\hat{\mathbf{p}}$:

$$\hat{A} = \sqrt{\frac{2E_p}{E_0 + E_p}}, \quad \hat{R} = \frac{c\hat{\mathbf{p}}\hat{\boldsymbol{\alpha}}}{E_0 + E_p} \tag{5.3}$$

with the consequence that the field-free Hamiltonian becomes diagonal,

$$\hat{U}_0 = \hat{A}(1 + \hat{R}\hat{\beta}) \quad \Rightarrow \quad \hat{U}_0^\dagger \hat{H}_0 \hat{U}_0 = E_p\hat{\beta} \tag{5.4}$$

and that this approach could be used to produce perturbation expansions to arbitrary order in the field $\hat{V}$. The resulting transformed potential terms can be subdivided into spin-free and spin–orbit terms. In a common approach, these are used separately.

The spin-free, or scalar, DK transformation is then used when computing the conventional one-electron integrals. The scalar part of the DKH Hamiltonian replaces the one-electron nonrelativistic Hamiltonian and all methods that are used in nonrelativistic calculations will automatically include these effects. Since the orbitals are generally different from the nonrelativistic ones, specific basis sets have been developed that are optimized for use with the DKH Hamiltonian. For the construction and results of these basis sets, see Chapter 6. The DKH transformation is iterative, and can be done to arbitrary order [12, 13].

The transformation amounts to a change of the basis set, and therefore integrals used to compute properties such as dipole moments must be computed in the new basis. This is known as a picture change and is most easily done for electric-field-like properties. In particular, for calculating electric field gradients, it is important that picture change effects are included [14].

Spin–orbit interaction can be included at any of several steps of a quantum chemistry calculation. If this is done together with orbital optimization, as in any DFT, SCF, or MCSCF calculation, two-component orbitals are used. These are general spin-orbitals, where each orbital contains contributions with α as well as β spin. (Obviously, so-called "unrestricted" orbitals cannot be used if spin–orbit interaction is to be calculated). In a four-component calculation, spin–orbit interaction is of course an inherent effect of the relativistic Hamiltonian and needs no separate treatment. The same is true for two-component calculations if scalar versus tensorial components have not been separated; this is often done for convenience anyway but is not necessary. More often, orbitals are spin-restricted, and a scalar-relativistic calculation is used. The spin–orbit interaction is then added in a subsequent step, for example, together with a correlated calculation, either perturbative, coupled cluster, or of CI type. A very convenient choice is the so-called Atomic Mean Field Integrals (AMFI [15]), where the two-electron part of the spin–orbit interaction has been contracted with an atomic density matrix, resulting in an effective one-electron spin–orbit interaction operator, already when integrals are computed, rather than being part of a self-consistency procedure.

Calculations made by the authors, reported in this book, are done using the program package MOLCAS [16]. Relativity is then routinely included as scalar relativity by an arbitrary-order DKH transformation [17], using the relativistic all-electron basis set denoted ANO-RCC [18–22]. The two-electron integrals are used untransformed. Picture change of properties, and AMFI spin–orbit interaction integrals [15], are used when relevant for the study. The spin–orbit interaction is then done after orbitals have been determined in a CASSCF calculation, and after "dressing" the Hamiltonian with dynamic correlation by perturbation, using CASPT2 (see Chapter 12), in a final so-called state interaction calculation using the RASSI program (see Chapter 10).

5.4   FURTHER READING

We can recommend the following textbooks:

- K.G. Dyall and K. Faegri, Jr., "Introduction to Relativistic Quantum Chemistry", Oxford University Press, Inc., New York, 2007.
- U. Kaldor and S. Wilson, "Theoretical Chemistry and Physics of Heavy and Superheavy Elements", Kluwer Academic Publishers, Dordrecht, 2003.
- P. Pyykkö, "Relativistic effects in structural chemistry", Chem Rev 88, 563 (1988).
- M. Reiher and A. Wolf, "Relativistic Quantum Chemistry: The Fundamental Theory of Molecular Science", Wiley-VCH, Weinheim, 2009.

5.5   REFERENCES

[1] Pyykkö P. Relativistic effects in structural chemistry. Chem Rev 1988; 88: 563–594.
[2] Dyall KG, Faegri K Jr. Introduction to Relativistic Quantum Chemistry. New York: Oxford University Press, Inc.; 2007.
[3] Saue T, Faegri K, Gropen O. Relativistic effects on the bonding of heavy and superheavy hydrogen halides. Chem Phys Lett 1996; 263: 360–366.

[4] Sucher J. Foundations of the relativistic theory of many-electron atoms. Phys Rev A 1980; 22: 348–362.

[5] Douglas N, Kroll NM. Quantum electrodynamical corrections to the fine structure of helium. Ann Phys 1974; 82: 89–155.

[6] Hess BA. Applicability of the no-pair equation with free-particle projection operators to atomic and molecular structure calculations. Phys Rev A 1985; 32: 756–763.

[7] Hess BA. Relativistic electronic-structure calculations employing a two-component no-pair formalism with external-field projection operators. Phys Rev A 1986; 33: 3742–3748.

[8] Chang C, Pélissier M, Durand P. Regular two-component Pauli-like effective hamiltonians in Dirac theory. Phys Scr 1986; 34: 394–404.

[9] Rutkowski A. Relativistic perturbation theory 3. A new perturbation approach to the 2-electron Dirac-Coulomb equation. J Phys B 1986; 19: 3443–3455.

[10] Kutzelnigg W. Perturbation theory of relativistic corrections 2. Analysis and classification of known and other possible methods. Z Phys 1990; 15: 27–50.

[11] Reiher M, Wolf A. Relativistic Quantum Chemistry: The Fundamental Theory of Molecular Science. Weinheim: Wiley-VCH; 2009.

[12] Reiher M, Wolf A. Exact decoupling of the Dirac Hamiltonian. II. The generalized Douglas–Kroll–Hess transformation up to arbitrary order. J Chem Phys 2004; 121: 10945–10956.

[13] Reiher M. Douglas–Kroll–Hess theory: a relativistic electrons-only theory for chemistry. Theor Chem Acc 2006; 116: 241–252.

[14] Mastalerz R, Barone G, Lindh R, Reiher M. Analytic high-order Douglas-Kroll-Hess electric field gradients. J Chem Phys 2007; 127: Art. 074105.

[15] Schimmelpfennig B. AMFI, An Atomic Mean-Field Spin-Orbit Integral Program. Stockholm: Stockholm University; 1996.

[16] Aquilante F, De Vico L, Ferré N, Ghigo G, Malmqvist PÅ, Neogrády P, Pedersen TB, Pitoňák M, Reiher M, Roos BO, Serrano-Andrés L, Urban M, Veryazov V, Lindh R. Software news and update MOLCAS 7: the next generation. J Comput Chem 2010; 31: 224–247.

[17] Peng D, Reiher M. Exact decoupling of the relativistic Fock operator. Theor Chem Acc 2012; 131: 1081–1100.

[18] Roos BO, Veryazov V, Widmark PO. Relativistic ANO type basis sets for the alkaline and alkaline earth atoms applied to the ground state potentials for the corresponding dimers. Theor Chem Acc 2004; 111: 345.

[19] Roos BO, Lindh R, Malmqvist PÅ, Veryazov V, Widmark PO. Main group atoms and dimers studied with a new relativistic ANO basis set. J Phys Chem A 2004; 108: 2851–2858.

[20] Roos BO, Lindh R, Malmqvist PÅ, Veryazov V, Widmark PO. New relativistic ANO basis sets for transition metal atoms. J Phys Chem A 2005; 109: 6575–6579.

[21] Roos BO, Lindh R, Malmqvist PÅ, Veryazov V, Widmark PO. New relativistic ANO basis sets for actinide atoms. Chem Phys Lett 2005; 409: 295–299.

[22] Roos BO, Lindh R, Malmqvist PÅ, Veryazov V, Widmark PO, Borin AC. New relativistic ANO basis sets for lanthanide atoms with applications to the Ce diatom and LuF_3. J Phys Chem A 2008; 112: 11431–11435.

6

BASIS SETS

In Chapter 3, we introduced the concepts of MO theory and in its base form, we assume that pure atomic orbitals combine and form molecular orbitals, the so-called linear combination of atomic orbitals approach or LCAO for short. For example, we describe the bonding in H_2 by combining the two 1s atomic orbitals into one bonding and one antibonding molecular orbital.

When we do real quantum chemical calculations, we need to go beyond this basic LCAO approach and add more functions with which we describe molecular orbitals. Such functions are normally atom-centered and we refer to them as atomic orbitals or atomic basis functions, or as basis sets.

The requirements of such basis sets largely depend on two factors: the accuracy aimed for and the methods used. Results for methods that only depend on occupied orbitals such as Hartree–Fock and Density Functional Theory converge quickly with the size of the basis set, whereas for implicitly correlated methods such as Full CI, see Section 2.3, such convergence is much slower.

Most basis sets published are intended for valence chemistry and are not suitable for studying Rydberg states, Mössbauer spectroscopy, core hole excitations, etc.

This chapter describes general concepts about basis sets, how to construct them, and how to select a basis set.

6.1 GENERAL CONCEPTS

When computing the molecular orbitals for the hydrogen molecule, we can expand these in a precomputed set of functions: basis functions or a basis set. As the simplest

Multiconfigurational Quantum Chemistry, First Edition.
Björn O. Roos, Roland Lindh, Per Åke Malmqvist, Valera Veryazov, and Per-Olof Widmark.
© 2016 John Wiley & Sons, Inc. Published 2016 by John Wiley & Sons, Inc.

basis set we have two functions, one centered on each nucleus. At large separation, these would be best described by pure hydrogenic 1s orbitals, which are proportional to e^{-r/a_0}, where r is the nucleus–electron distance.

If we optimize the structure of H_2 using the Hartree–Fock method with this basis, we get an interaction that is much too weak, the bond length is too long, and the bond strength is too low. We can introduce a variational parameter in our basis and let the basis function be proportional to $e^{-\zeta r/a_0}$, where we optimize the parameter ζ. Near equilibrium, the optimal value for ζ is close to 1.25. Let us use one function on each atom with $\zeta = 1.25$ as basis functions, denote them 1s', and optimize the structure of H_2, we then get a very short bond with a reasonable energy; see Table 6.1. Using both 1s and 1s' on each atom gives a bond distance close to the Hartree–Fock limit, while the bond strength is still too low.

It is only with the introduction of basis functions that allows for anisotropic deformation of the spherical shape of the atomic orbitals that we get results that approach what is possible with the Hartree–Fock method. This can be accomplished with a p-function centered on each atom. The first spectroscopic p-orbital is proportional to $re^{-0.5r/a_0}$. If we, on the other hand, optimize the exponent to get the lowest possible Hartree–Fock energy at the equilibrium distance the exponent is about 3.8. The best possible p-function is thus proportional to $re^{-3.8r/a_0}$, quite different from the spectroscopic 2p orbital. Let us call this optimized function 1p. In Table 6.1, we compare the results for a few combinations of 1s, 1s', and 1p basis functions, and the Hartree–Fock limit realized by using a saturated Gaussian basis set.

By using a basis set with two s-functions with different values for ζ (1 and 1.25) plus a p-function to polarize the s-functions, we get a reasonably accurate result. A basis set of this size is generally referred to as a Double ζ plus Polarization basis or DZP basis for short. A general conclusion is that we need at least a DZP basis to obtain quantitative results.

See below for a more detailed explanation of these acronyms.

6.2   SLATER TYPE ORBITALS, STOs

When we solve the nonrelativistic Schrödinger equation for hydrogenic atoms with a nucleus that is treated as a point charge, we get solutions that in the radial part is

TABLE 6.1   Hartree–Fock Structures and Energies for H_2 with Various Basis Sets

Basis	R[Å]	E(au)
1s	0.848	−1.0991
1s'	0.706	−1.1257
1s + 1s'	0.733	−1.1283
1s + 1s' + 1p	0.734	−1.1334
HF-limit	0.734	−1.1336

an exponential function multiplied with a polynomial. The $1s$ orbital is particularly simple, $\varphi_{1s} = Ne^{-r/a_0}$, a simple exponential function. It is tempting to use such functions as basis functions; they yield fairly accurate results with relatively few functions. Basis sets of this type is referred to as Slater Type Orbitals, STO for short.

There are a few problems with STOs. The first problems are of technical nature; among other things, it is hard to compute two-electron integrals for more than two centers [1, 2].

Slater type basis sets suffer from two other problems of more fundamental nature. In real life, the nuclei are not point charges but have a radius in the order of femtometers. Thus, the orbitals should not have a cusp at the nucleus that the STOs all have. The effect is small and is not really noticeable for ordinary valence chemistry, but the effect can be significant if you are studying properties that depend on the electronic amplitude at the nuclei such as, for example, Mössbauer spectroscopy. Another problem with STO basis sets is that you typically need small exponents to get a good description with relatively few functions. These small exponents lead to a gross overestimate of the interaction at very long distances. Again not a very serious problem unless you really are interested in the interactions at very long distances.

6.3   GAUSSIAN TYPE ORBITALS, GTOs

One way to circumvent the problems with STOs is to expand the Slater function in a sum of Gaussian functions such as

$$\varphi_{\text{Slater}} \approx \sum_k c_k e^{-\alpha_k r^2}, \tag{6.1}$$

where the α's are called exponents and the c's are called contraction coefficients, and $e^{-\alpha_k r^2}$ is called a primitive Gaussian basis function. The reason such a rewriting is beneficial is that the computation of two-electron integrals with Gaussians is far more easy to perform.

A natural idea is not to use STOs at all, just to use Gaussian functions as basis functions. We call such functions Gaussian type orbitals, GTO. Sometimes, they are referred to as Gaussian type functions, GTF. There is one major problem with using GTOs as basis functions; we need many Gaussian functions to get the shape of the orbitals correct and the size of the basis set quickly becomes too large for practical calculations. A solution is to do a contraction in the same way as we do when expanding an STO in GTOs. Such basis functions are called contracted Gaussian type orbitals, CGTO, and is the dominating type of basis sets used today. There are a few different strategies to determine the contraction coefficients, described as follows.

The remainder of this chapter is about Gaussian basis set design.

6.3.1   Shell Structure Organization

Basis sets are in general organized in the same way as the atomic orbitals are, we have shells, the 1s orbital is described by one or more functions, the 2s orbital by one or more functions and so on. When we come to the 2p orbital, the three components $2p_x$, $2p_y$, and $2p_z$ are all described by the same radial function(s), and similarly for higher angular momentum.

6.3.2   Cartesian and Real Spherical Harmonics Angular Momentum Functions

The general form for a primitive basis function is $N_l(\alpha_k)r^l e^{-\alpha_k r^2} Y_{l,m}(\theta, \phi)$, where $Y_{l,m}$ are spherical harmonics, l specifies the magnitude of the angular momentum, and m specifies the z-component of the angular momentum. The normalization constant $N_l(\alpha)$ depends on both the subshell and the exponent.

The spherical harmonics are complex valued functions; but in chemistry, we typically want to deal with real-valued orbitals. We can construct real-valued functions, so-called real spherical harmonics, by making linear combinations; for example, a p_x function can be made with the combination $Y_{1,1} + Y_{1,-1}$. We can represent all angular parts of the orbitals with Cartesian representations, $x^{k_x} y^{k_y} z^{k_z}$, where the sum of the exponents is the angular momentum quantum number, that is, $k_x + k_y + k_z = l$.

For $l = 2$, there are six possible Cartesian terms, x^2, xy, xz, y^2, yz, and z^2, but there are only five d components. One linear combination, $x^2 + y^2 + z^2$, presents a spherically symmetric function and is hence an s-function. In a similar manner, we have 10 Cartesian terms for $l = 3$ and 3 linear combinations correspond to p functions, while the remaining 7 are f orbitals. For $l = 4$, we get one s component, five d as well as the nine proper g components.

Some basis sets have been constructed with all Cartesian terms present, for example, the 6-31G family [3, 4] and it is important to use all components in calculations using these basis sets.

There are some issues with using all Cartesian components in calculations and most modern basis sets have been constructed with five real d components, seven real f components etc., using so-called real spherical harmonics.

It is very important to use the basis sets as designed in molecular calculations. Using all Cartesian components with a basis set designed with real spherical harmonics will almost certainly lead to problems associated with near-linear dependence such as poor convergence and numerical instability. On the other hand, using just the real spherical harmonics with a basis set designed with all Cartesian components will lead to a severe deficiency in the basis set, probably giving large basis set superposition errors (BSSE).

6.4   CONSTRUCTING BASIS SETS

In molecular calculations, we almost invariably use precomputed atom-centered basis sets, that is, they contain fixed functions in which we expand our molecular orbitals. The first question that arises when we start to design a basis set is what criteria to use for the design. Most would agree with three somewhat incompatible requirements:

1. The basis set should represent the atoms well, that is, be close to the complete basis set limit.

2. The basis set should be flexible enough to be able to describe all distortions of the atom in various situations.

3. The basis set should be small in all respects to minimize the computational effort.

The third criterion is clearly in conflict with the first two and all basis sets that are available in the literature are compromises. Many of the basis sets available are constructed to give good results for ordinary ground state chemistry, bond breaking, etc. These general-purpose basis sets do not, however, handle things such as, for example, Rydberg orbitals (see below) properly; here we need special-purpose basis sets or augment the general-purpose basis sets properly. There are quite a few general-purpose basis sets in the literature. They tend to have slightly different design criteria and it is a very good idea to read the original article, where the basis set is first published.

6.4.1 Obtaining Exponents

The very first thing to do is to obtain the exponents (α_k) used to define the primitive Gaussian basis functions $\chi_k = N_k \, x^{k_x} y^{k_y} z^{k_z} \, e^{-\alpha_k r^2}$.

These exponents are typically determined by minimizing the Hartree–Fock energy for the atom in its ground state. An energy minimization is in general a good criterion but will not necessarily yield the flexibility needed for certain types of calculations. For example, the total energy is relatively insensitive to the detailed shape of the $1s$ orbital close to the nuclear cusp. A general-purpose basis set would, for example, have to be augmented for calculations that measures the electronic amplitude at the nucleus.

Using the Hartree–Fock energy when optimizing exponents only gives functions of the occupied shells; the Hartree–Fock energy is invariant to the shape of the virtual orbitals. It might be tempting to use excited states where the next subshells are occupied, but this leads to functions that are much to diffuse to be useful in molecular calculations. One method used to obtain these, so-called polarization functions, is to match the maximum of the radial distribution function with the function that is to be polarized. Today, it is fairly common to determine the exponents by minimizing the correlation energy using some correlated method such as MP2, for which the energy depends on the shape of all orbitals including the virtual orbitals.

6.4.2 Contraction Schemes

After the exponents have been determined, those are fixed and the associated contraction coefficients are determined.

Perhaps the most common approach to determine the contraction coefficients is to use the Hartree–Fock method, where each atomic orbital, φ_i, is expanded in the primitive Gaussian basis functions

$$\varphi_i = \sum_k c_{k,i} \chi_k. \tag{6.2}$$

The contraction coefficients are determined variationally in the SCF procedure. This gives a description of each occupied orbital in the ground state that can then be used as CGTOs in molecular calculations. There are also virtual orbitals defined in the SCF procedure, but these orbitals have again little value for basis set design.

If you use the SCF orbitals to determine the contraction coefficients in such a way that all primitive s basis functions are used to describe the 1s basis function with the corresponding coefficients, using the coefficients from the 2s orbital for the 2s basis function, etc., you will obtain a CGTO basis set that will be able to represent the atomic orbitals exactly, within the subspace defined by the primitive basis set.

This is the best you can do in terms of contracting the basis set as long as you only are interested in the SCF solution for the atom. Such a minimal contraction would be termed Minimal Basis, MB, and is in most cases of no practical use, except for preliminary calculations. When the environment around an atom changes, such as in a molecule, the atomic orbitals will be deformed and we need to include other basis functions to facilitate this. We clearly need a way to introduce some flexibility and this can be done in different ways and is discussed as follows.

There are two main types of contractions schemes: segmented contraction and general contraction, and a number of schemes in between. In general contraction, all primitive basis functions contribute to all contracted basis functions similar to the discussion in the paragraph above. In segmented contraction, each primitive basis function contributes to only one contracted function, which may seem less than optimal. There are some technical advantages with segmented contraction, however. In Figure 6.1, we see an example of how 11 primitive s-functions can be contracted for sodium, which have 1s–3s occupied. For the general contraction, each primitive function contributes to each of the 1s–3s atomic orbitals. For the segmented contraction scheme, we have a Double Zeta, DZ, contraction with two contracted functions for each atomic orbital.

The pedestrian way of making a, for example, segmented DZ contraction is to take the canonical orbitals from an atomic SCF calculation and look at the coefficients for

	General			Segmented						Overlapping			
	1s	2s	3s	1s	1s′	2s	2s′	3s	3s′	1s	2s	3s	3s′
χ_1	■	■	■	■	0	0	0	0	0	■	0	0	0
χ_2	■	■	■	■	0	0	0	0	0	■	0	0	0
χ_3	■	■	■	■	0	0	0	0	0	■	0	0	0
χ_4	■	■	■	■	0	0	0	0	0	■	0	0	0
χ_5	■	■	■	■	0	0	0	0	0	■	0	0	0
χ_6	■	■	■	0	■	0	0	0	0	■	■	0	0
χ_7	■	■	■	0	■	0	0	0	0	0	■	0	0
χ_8	■	■	■	0	0	■	0	0	0	0	■	■	0
χ_9	■	■	■	0	0	0	■	0	0	0	■	■	0
χ_{10}	■	■	■	0	0	0	0	■	0	0	0	■	0
χ_{11}	■	■	■	0	0	0	0	0	■	0	0	0	■

Figure 6.1 General versus segmented versus overlapping contraction.

the 1s orbital and take all that are significantly different from zero and use them for the 1s and 1s' and make a split where the coefficient are the largest. For the 2s and 2s' you look at the coefficients for the 2s orbital and take all that are significantly different from zero of the remaining primitives and make a split in the same manner. The procedure is repeated for the remaining atomic orbitals.

The use of two functions for the atomic 1s in sodium is wasteful since the 1s do not participate in normal chemistry and remains inert. Still we need to have the 1s split in two parts to have a reasonable accuracy for the 2s and 3s. Various ways have been devised to cope with this and one is overlapping contraction illustrated in Figure 6.1.

The terminology used in the literature is sometimes somewhat inconsistent and confusing. A proper Double Zeta contraction should contain two functions for each atomic orbital, including the core orbitals. The overlapping contraction in Figure 6.1 represents a contraction that usually is referred to as Split Valence contraction, SV. A terminology that perhaps is more commonly used today is Valence Double Zeta contraction, VDZ, rather than Split Valence contraction. The acronym VDZ indicates that the basis set has a DZ contraction for the valence only. For example, a VDZ basis set for carbon would contain one function to describe the $1s$ atomic orbital, two functions to describe the 2s atomic orbital, and two functions to describe the 2p atomic orbital.

As we saw for the hydrogen molecule above, we needed a p-function to get a quantitatively correct result. Such a function is referred to as a polarization function and a basis set for hydrogen that contains two s-functions and one p-function is referred to as a double zeta plus polarization, DZP, basis set. We can of course increase the size of the basis set to three s-functions, two p-functions, and one d-function yielding a basis set of triple zeta plus double polarization, TZ2P, quality. Today, the "Double" is commonly omitted and simply referred to as TZP. The notation is continued with Quadruple Zeta, QZ, for four functions and Quintuple Zeta, 5Z, for five functions, etc. Sometimes, you see the acronyms in another order such as Polarized Valence Double Zeta, pVDZ.

Segmented contraction is fairly straightforward; you decide which primitive functions to assign to which atomic orbitals. The contraction coefficients are usually derived from SCF orbitals for the atom. Ahlrichs et al. [5–11] have published a sequence of fully optimized segmented basis sets. They optimize all parameters in the basis set with respect to the Hartree–Fock energy of the atoms, with the contraction pattern as only constraint.

General contraction is, on the other hand, less obvious. One easy way is to use the SCF coefficients as contraction coefficients and decontract the most diffuse functions. This is referred to as a Raffenetti contraction [12] and, continuing our example with the contraction of the s-functions of sodium, a VTZ contraction is shown in Figure 6.2. One popular family of basis sets that uses Raffenetti contraction is the correlation-consistent basis sets [13–35]. Another way of obtaining the contraction coefficients is the Atomic Natural Orbital approach introduced by Almlöf and Taylor [36, 37]. The contraction coefficients are taken from the natural orbitals of an implicitly correlated calculation on the atom. The orbitals with high occupation numbers are used in the basis set while those with small occupation numbers are discarded.

	1s	2s	3s	3s′	3s″
χ_1	■	■	■	0′	0
χ_2	■	■	■	0	0
χ_3	■	■	■	0	0
χ_4	■	■	■	0	0
χ_5	■	■	■	0	0
χ_6	■	■	■	0	0
χ_7	■	■	■	0	0
χ_8	■	■	■	0	0
χ_9	■	■	■	0	1
χ_{10}	■	■	■	0	0
χ_{11}	■	■	■	1	0

Figure 6.2   Raffenetti contraction.

If only the valence orbitals are correlated we get basis sets of quality VDZ, VTZ, etc. With this approach, the coefficients for the polarization functions are also well defined.

There is here a bias toward the atomic state used in the contraction and for atoms, where more than one atomic state is involved in the chemistry, for example, transition metals, this can be problematic. One way to resolve this is to include more than one atomic state in determining the contraction coefficients as in the Density Matrix Averaged Atomic Natural Orbital approach [38–46]. Rather than diagonalizing the density matrix from one state, several density matrices are averaged and you will treat all included atomic states in a more balanced way. Different emphasis can be put on different states by using different weights but using the same weight on all the states included will normally yield a balanced basis set.

Let us look at the importance of considering several states when contracting a basis set. Consider the Ni atom that has a $^3F(3d^84s^2)$ ground state that is almost degenerate with the $^3D(3d^94s)$ state. It is important to have a balanced description of these two states to be able to describe the chemistry of nickel. In Table 6.2 is shown a sequence of SCF energies for the two states using different contractions. The nonrelativistic SCF excitation energy with a 21s15p10d primitive basis set is 1.277 eV and will serve as reference. If we take the SCF orbitals from an atomic calculation on the ^3F state as contraction coefficients, this state will be perfectly reproduced with a 4s2p1d basis set, one basis function per occupied orbital. The ^3D state, on the other hand, is not well represented and the excitation energy increases to 5.365, a truncation error of about 4eV. If we do the reverse and base the contraction on the ^3D state, we get an equally bad truncation error, about 4eV in the other direction. Averaging the density matrix of the two states, we obtain a 6s3p2d contraction, and we will reproduce both states with the same excitation energy as for the calculation with the primitive basis set.

It might be tempting to assume that the flexibility of the basis set describes the different shapes of the 3d and 4s orbitals with the 3s and 3p orbitals inert. If we redo the averaging but only use the 3s orbital from the ^3F state, we do get a 5s3p2d contraction, one less s-basis function. Clearly the basis set includes flexibility to allow

TABLE 6.2   SCF Energies for the Ni Atoms with Respect to Different Contraction Schemes

Size	Contraction	$E(^3F)$(au)	$E(^3D)$(au)	ΔE(eV)
21s15p10d	Primitive	−1506.870252	−1506.823332	1.277
4s2p1d	^3F	−1506.870252	−1506.673106	5.365
4s2p1d	^3D	−1506.722825	−1506.823332	−2.735
6s3p2d	$^3F+^3D$	−1506.870251	−1506.823331	1.277
5s3p2d	$^3F+^3D^a$	−1506.870251	−1506.818935	1.396
6s5p3d	cc-pVDZ	−1506.867450	−1506.822248	1.230
7s6p4d	cc-pVTZ	−1506.869252	−1506.822293	1.278

[a]Shape of 3s only comes from the ^3F state.

TABLE 6.3   SDCI Energies for Ni Atom with Respect to Different Contractions Sizes for an ANO Basis with Bias Toward the ^3F State

Size	$E(^3F)$(au)	$E(^3D)$(au)	ΔE(eV)
21s15p10d[a]	−1507.05877302	−1507.04278610	0.435
5d4p3d	−1507.04858937	−1507.02936322	0.523
6d5p4d	−1507.05612516	−1507.03956925	0.451
7d6p5d	−1507.05791192	−1507.04160918	0.444
8d7p6d	−1507.05843714	−1507.04230333	0.439
9d8p7d	−1507.05868448	−1507.04267955	0.436

[a]Uncontracted.

the 3s to relax, and it is clear from Table 6.2 that this effect is important, with about 0.1 eV truncation error if the flexibility of the 3s is left out of the basis set. In Table 6.2, you also see the results for the standard cc-pVDZ and cc-pVTZ [30] basis sets for comparison.

A basis set of ANO type where the contraction coefficients come from a correlated calculation of one states will yield a reasonable representation of the other state if a sufficient number of basis functions is used. In Table 6.3, you see the convergence of the total singles and double CI (SDCI) energy and relative energy of the two states. To get a truncation error in the excitation of less than 0.04 eV (chemical accuracy), we need a basis set of size 7s6p5d, a QZ contraction.

6.4.3   Convergence in the Basis Set Size

For any variational method such as SCF or Full CI, a systematic extension of the size of the basis set will lead to a decrease of the energy, which will converge to the complete basis set limit of the method. The rate of this convergence depends very much on the nature of the calculation. The wave function of an SCF or MCSCF calculation depends on the shape of a fixed number of orbitals, the number of occupied orbitals for the SCF calculation, and the number of inactive and active orbitals in the second case. The wave function for an MRCI calculation, on the other hand, depends (in principle at least) on the shape of all orbitals including all virtual orbitals. The rate of convergence with basis set size is relatively fast if the wave function depends on the shape of a fixed number of orbitals as for SCF and MCSCF but is very slow for methods that depend on all orbitals such as MRCI. The reason for the very slow convergence in the latter case is the fact that the wave function contains a cusp in the electron–electron distance for each electron–electron pair and a Gaussian type basis sets only contains smooth functions; Kutzelnigg [47] made a thorough study of the convergence pattern.

The oxygen atom has a ground state O $^3P(2s^22p^4)$ with the lowest excited state O $^1D(2s^22p^4)$ and for the anion the ground state is O$^-$ $^2P(2s^22p^5)$.

In Tables 6.4 and 6.5, the convergence of the energy of these levels with respect to basis set size is shown. The basis set used is the ANO-RCC [42–46] with DZP, TZP, and QZP contraction, and a Douglas–Kroll–Hess Hamiltonian to second order. The SCF results are more or less converged for the TZP contraction with respect to both total energies and energy differences. For the SDCI energies, the total energies are not converged and neither for the anion, but the excitation energy is almost. The main difference here between O $^1D(2s^22p^4)$ and O$^-$ $^2P(2s^22p^5)$ is that in the latter case the number of electrons is changed with respect to the ground state while in the former case the number of electrons remains unchanged. Feller and Davidson [49] made a thorough study of the electron affinity of oxygen. In general, if there is a substantial rearrangement of the electronic structure, you get a large differential electron–electron correlation energy. In such cases, the convergence in basis set size is slow and large basis sets are needed.

TABLE 6.4 SCF Energies for the O $^3P(2s^22p^4)$, O $^1D(2s^22p^4)$ and O$^-$ $^2P(2s^22p^5)$ Levels

Size	$E(^3P)$	$E(^1D)$	$E(^2P)$	$\Delta E(^1D)$	$\Delta E(^2P)$
3s2p1d	−74.860183	−74.780322	−74.837120	2.173	−0.628
4s3p2d1f	−74.861287	−74.781002	−74.841367	2.185	−0.542
5s4p3d2f1g	−74.861499	−74.781365	−74.841566	2.181	−0.542

Note: Absolute energies are in au while energy difference are in eV.

TABLE 6.5 SDCI Energies for the O $^3P(2s^22p^4)$, O $^1D(2s^22p^4)$ and O$^-$ $^2P(2s^22p^5)$ Levels

Size	$E(^3P)$	$E(^1D)$	$E(^2P)$	$\Delta E(^1D)$	$\Delta E(^2P)$
3s2p1d	−74.98232676	−74.90334262	−75.00414227	2.149	0.594
4s3p2d1f	−75.02592420	−74.95106198	−75.06135883	2.037	0.964
5s4p3d2f1g	−75.03895997	−74.96556887	−75.07708277	1.997	1.037
Expt.				1.968[a]	1.46[b]

Note: Absolute energies are in au while energy difference are in eV.
[a]Excitation energy is taken from www.nist.gov.
[b]Reference [48]

6.5   SELECTION OF BASIS SETS

If you want to assess the suitability of the basis set you intend to use in your calculations, you need to consider what are the requirements you have. If you include relativity, make sure that the basis sets are for relativistic calculations; if you need to correlate semicore orbitals, make sure that the basis set include such effects, etc. The ultimate guide is the original article when the basis set was published; there you will find design criteria. Review articles may be useful, but unfortunately you occasionally see reviewers that have not done their homework by really reading the original articles.

6.5.1   Effect of the Hamiltonian

It is very important to use a basis set constructed for the Hamiltonian you are using in your calculations. If we use a contracted basis set constructed for a nonrelativistic Hamiltonian in a relativistic calculation, we do get very large truncation errors. The same problem is seen with a relativistic basis set used in nonrelativistic calculations. The effect of using a nonrelativistic ANO basis set in a Douglas–Kroll–Hess calculation at the SCF level is shown in Table 6.6. The truncation error does not converge in any reasonable way and is very large even for the relatively light atom nickel ($Z = 28$). In Table 6.7, the same poor convergence pattern is shown for SDCI calculations. However, the excitation energy is not that bad and it might be tempting to go along and perform such calculation, but this might lead to very strange results. The deficiency is about 80 eV, a very large error, and any basis function available on some other atom might be used to improve the orbitals resulting in very large basis set superposition errors.

Using the Cl atom ($Z = 17$) as another example is shown in Figure 6.3, where basis sets with a contraction for nonrelativistic calculations have been used with a DKH Hamiltonian to second order. The error is relative to the energy with the primitive 17s11p basis set. The convergence is very slow and within the limit of a valence quadruple zeta (vQZ) contraction, no real improvement can be seen. Two different ways of constructing the contractions have been chosen: an ANO contraction using

TABLE 6.6   SCF Energies for Ni Atom Using a Nonrelativistic Contraction with a Douglas–Kroll–Hess Hamiltonian to Second Order

Size	$E(^3F)$(au)	$E(^3D)$(au)	ΔE(eV)
21s15p10d[a]	−1519.07008708	−1519.00983217	1.640
5d4p3d	−1516.33204460	−1516.27016749	1.684
6d5p4d	−1516.34036182	−1516.28000929	1.642
7d6p5d	−1516.34917879	−1516.28948326	1.624
8d7p6d	−1516.35994243	−1516.30059874	1.615
9d8p7d	−1516.38334516	−1516.32412018	1.612

[a]Uncontracted.

TABLE 6.7   SDCI Energies for Ni Atom Using a Nonrelativistic Contraction with a Douglas–Kroll–Hess Hamiltonian to Second Order

Size	$E(^3F)$(au)	$E(^3D)$(au)	ΔE(eV)
21s15p10d[a]	−1519.26093331	−1519.23181391	0.792
5d4p3d	−1516.51282768	−1516.48204826	0.838
6d5p4d	−1516.52873671	−1516.49960976	0.793
7d6p5d	−1516.53937207	−1516.51076420	0.778
8d7p6d	−1516.55046800	−1516.52219122	0.769
9d8p7d	−1516.57419485	−1516.54600478	0.767

[a]Uncontracted.

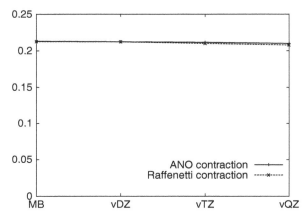

Figure 6.3   Contraction errors, in Hartrees, for Cl atom using nonrelativistic basis sets with a DKH Hamiltonian to second order.

the density from an SDCI calculation on the atom in its ground state and a Raffenetti contraction based on an SCF calculation on the atom in its ground state.

6.5.2   Core Correlation

The core orbitals of the lighter elements, such as the 1s of oxygen, are inert and unimportant to include in any correlation treatment, unless you require extremely high accuracy. As we go down the periods in the periodic table, the difference in energy between outermost core, semicore, and the valence gets smaller. This is particularly noticeable for the first groups such as the alkali atoms and alkali earth metals. Contracted basis sets used in calculations with the semicore correlated need to include basis functions that describe such correlation. For basis sets of ANO type, atomic densities with the semicore correlated need to be included. For basis sets of Raffenetti type, primitives in the range that describe the semicore need to be decontracted.

6.5.3 Other Issues

Most published basis sets are designed to reproduce the core and valence orbitals, for example, 1s–4s, 2p–3p, and 3d for the nickel atom. In some more extensive basis sets, there would be a description of the 4p as well. The atomic orbitals with main quantum number larger than that of the highest occupied orbital are called Rydberg orbitals and are much more diffuse than the valence orbitals and there are no basis functions to describe these in a standard basis set. For example, the 5s of nickel is not included in any standard basis set and any attempt to perform a calculation where this orbital contributes to the electronic structure of the molecule would yield very poor results. To perform calculation on Rydberg states, the basis sets need to be augmented; see for example [50].

Other properties not well represented by standard basis sets is at the other end of the scale, properties that depend on the amplitude of the orbitals at a nucleus. For such calculations, you need to augment the basis sets with compact functions; see for example Ref. [51].

6.6 REFERENCES

[1] Boys SF. Electronic wave functions. I A general method of calculation for the stationary states of any molecular system. Proc R Soc London, Ser A 1950;200:542.

[2] Roothaan CCJ. A study of two-center integrals useful in calculations on molecular structure. I. J Chem Phys 1951;19:1445–1458.

[3] Hehre WJ, Ditchfield R, Pople JA. Self-consistent molecular orbital methods. XII. Further extensions of Gaussian type basis sets for use in molecular orbital studies of organic molecules. J Chem Phys 1972;56:2257.

[4] Petro WJ, Francl MM, Hehre WJ, DeFrees DJ, Pople JA, Binkley JS. Self-consistent molecular orbital methods. 24. Supplemented small split-valence basis sets for second-row elements. J Am Chem Soc 1982;104:5039–5048.

[5] Schäfer A, Horn H, Ahlrichs R. Fully optimized contracted Gaussian basis sets for atoms Li to Kr. J Chem Phys 1992;97:2571.

[6] Schäfer A, Huber C, Ahlrichs R. Fully optimized contracted Gaussian basis sets of triple zeta valence quality for atoms Li to Kr. J Chem Phys 1994;100:5829.

[7] Ahlrichs R, May K. Contracted all-electron Gaussian basis sets for Rb to Xe. Phys Chem Chem Phys 2000;2:943.

[8] Weigend F, Furche F, Ahlrichs R. Gaussian basis sets of quadruple zeta valence quality for atoms H–Kr. J Chem Phys 2003;119:12753.

[9] Weigend F, Ahlrichs R. Balanced basis sets of split valence, triple zeta valence and quadruple zeta valence quality for H to Rn: design and assessment of accuracy. Phys Chem Chem Phys 2005;7:3297.

[10] Rappoport D, Furche F. Property-optimized Gaussian basis sets for molecular response calculations. J Phys Chem 2010;133:134105.

[11] Weigend F, Baldes A. Segmented contracted basis sets for one- and two-component Dirac-Fock effective core potentials. J Chem Phys 2010;133:174102.

[12] Raffenetti RC. General contraction of Gaussian atomic orbitals: core, valence, polarization and diffuse basis sets; molecular integral evaluation. J Chem Phys 1973;58:4452–4458.

[13] Dunning TH Jr. Gaussian basis sets for use in correlated molecular calculations. I. The atoms boron through neon and hydrogen. J Chem Phys 1989;90:1007–1023.

[14] Kendall RA, Dunning TH Jr., Harrison RJ. Electron affinities of the first-row atoms revisited. Systematic basis sets and wave functions. J Chem Phys 1992;96:6796–6806.

[15] Woon DE, Dunning TH Jr. Gaussian basis sets for use in correlated molecular calculations. IV. Calculation of static electrical response properties. J Chem Phys 1994;100:2975–2988.

[16] Woon DE, Dunning TH Jr. Gaussian basis sets for use in correlated molecular calculations. V. Core-valence basis sets for boron through neon. J Chem Phys 1995;103:4572–4585.

[17] Peterson KA, Dunning TH Jr. Accurate correlation consistent basis sets for molecular core-valence correlation effects. The second row atoms Al–Ar, and the first row atoms B–Ne revisted. J Chem Phys 2002;117:10548.

[18] Wilson AK, van Mourik T, Dunning TH Jr. Gaussian basis sets for use in correlated molecular calculations. VI. Sextuple-zeta correlation-consistent sets for boron through neon. J Mol Struct THEOCHEM 1996;388:339–349.

[19] van Mourik T, Wilson AK, Dunning TH Jr. Benchmark calculations with correlated molecular wavefunctions. XIII. Potential energy curves for He_2, Ne_2 and Ar_2 using correlation consistent basis sets through augmented sextuple zeta. Mol Phys 1999;99:529–547.

[20] Woon DE, Dunning TH Jr. Gaussian basis sets for use in correlated molecular calculations. III. The second row atoms, Al–Ar. J Chem Phys 1993;98:1358–1371.

[21] van Mourik T, Dunning TH Jr. Gaussian basis sets for use in correlated molecular calculations. VIII. Standard and augmented sextuple zeta correlation consistent basis sets for aluminum through argon. Int J Quantum Chem 2000;76:205–221.

[22] Dunning TH Jr., Peterson KA, Wilson AK. Gaussian basis sets for use in correlated molecular calculations. X. The atoms aluminum through argon revisited. J Chem Phys 2001;114:9244–9253.

[23] Hill JG, Mazumder S, Peterson KA. Correlation consistent basis sets for molecular core-valence effects with explicitly correlated wave functions: the atoms B–Ne and Al–Ar. J Chem Phys 2010;132:054108.

[24] Wilson AK, Woon DE, Peterson KA, Dunning TH Jr. Gaussian basis sets for use in correlated molecular calculations. IX. The atoms gallium through krypton. J Chem Phys 1999;110:7667–7676.

[25] de Jong WA, Harrison RJ, Dixon DA. Parallel Douglas-Kroll energy and gradients in NWChem: estimating scalar relativistic effects using Douglas-Kroll contracted basis sets. J Chem Phys 2001;114:48–53.

[26] DeYonker NJ, Peterson KA, Wilson AK. Systematically convergent correlation consistent basis sets for molecular core-valence correlation effects: the third-row atoms gallium through krypton. J Phys Chem A 2007;111:11383.

[27] Peterson KA. Systematically convergent basis sets with relativistic pseudopotentials. I. Correlation consistent basis sets for the post-d group 13–15 elements. J Chem Phys 2003;119:11099.

[28] Peterson KA, Figgen D, Goll E, Stoll H, Dolg M. Systematically convergent basis sets with relativistic pseudopotentials. II. Small-core pseudopotentials and correlation consistent basis sets for the post-d group 16–18 elements. J Chem Phys 2003;119:11113.

[29] Peterson KA, Yousuf KE. Molecular core-valence correlation effects involving the post-d elements Ga–Rn: benchmarks and new pseudopotential-based correlation consistent basis sets. J Chem Phys 2010;133:174116.

[30] Balabanov NB, Peterson KA. Systematically convergent basis sets for transition metals. I. All-electron correlation consistent basis sets for the 3d elements Sc-Zn. J Chem Phys 2005;123:064107.

[31] Balabanov NB, Peterson KA. Basis set limit electronic excitation energies, ionization potentials, and electron affinities for the 3d transition metal atoms: coupled cluster and multireference methods. J Chem Phys 2006;125:074110.

[32] Peterson KA, Puzzarini C. Systematically convergent basis sets for transition metals. II. Pseudopotential-based correlation consistent basis sets for the group 11 (Cu, Ag, Au) and 12 (Zn, Cd, Hg) elements. Theor Chem Acc 2005;114:283.

[33] Peterson KA, Figgen D, Dolg M, Stoll H. Energy-consistent relativistic pseudopotentials and correlation consistent basis sets for the 4d elements Y–Pd. J Chem Phys 2007;126:124101.

[34] Figgen D, Peterson KA, Dolg M, Stoll H. Energy-consistent pseudopotentials and correlation consistent basis sets for the 5d elements Hf–Pt. J Chem Phys 2009;130:164108.

[35] Prascher BP, Woon DW, Peterson KA, Dunning TH Jr., Wilson AK. Gaussian basis sets for use in correlated molecular calculations. VII. Valence and core-valence basis sets for Li, Na, Be, and Mg. Theor Chem Acc 2011;128:69.

[36] Almlöf J, Taylor PR. General contraction of Gaussian basis sets. I. Atomic natural orbitals for first- and second-row atoms. J Chem Phys 1987;86:4070–4077.

[37] Almlöf J, Taylor PR. General contraction of Gaussian basis sets. II. Atomic natural orbitals and the calculation of atomic and molecular properties. J Chem Phys 1990;92:551–560.

[38] Widmark PO, Malmqvist PÅ, Roos BO. Density matrix averaged atomic natural orbital (ANO) basis sets for correlated molecular wave functions. I. First row atoms. Theor Chim Acta 1990;77:291–306.

[39] Widmark PO, Persson BJ, Roos BO. Density matrix averaged atomic natural orbital (ANO) basis sets for correlated molecular wave functions. II. Second row atoms. Theor Chim Acta 1991;79:419–432.

[40] Pou-Amérigo R, Merchán M, Nebot-Gil I, Widmark PO, Roos BO. Density matrix averaged atomic natural orbital (ANO) basis sets for correlated molecular wave functions. III. First row transition metal atoms. Theor Chim Acta 1995;92:149–181.

[41] Pierloot K, Dumez B, Widmark PO, Roos BO. Density matrix averaged atomic natural orbital (ANO) basis sets for correlated molecular wave functions. IV. Medium size basis sets for the atoms H–Kr. Theor Chim Acta 1995;92:87–114.

[42] Roos BO, Veryazov V, Widmark PO. Relativistic ANO type basis sets for the alkaline and alkaline earth atoms applied to the ground state potentials for the corresponding dimers. Theor Chem Acc 2004;111:345.

[43] Roos BO, Lindh R, Malmqvist PÅ, Veryazov V, Widmark PO. Main group atoms and dimers studied with a new relativistic ANO basis set. J Phys Chem A 2004;108:2851–2858.

[44] Roos BO, Lindh R, Malmqvist PÅ, Veryazov V, Widmark PO. New relativistic ANO basis sets for transition metal atoms. J Phys Chem A 2005;109:6575–6579.

[45] Roos BO, Lindh R, Malmqvist PÅ, Veryazov V, O-Widmark P. New relativistic ANO basis sets for actinide atoms. Chem Phys Lett 2005;409:295–299.

[46] Roos BO, Lindh R, Malmqvist PÅ, Veryazov V, Widmark PO, Borin AC. New relativistic ANO basis sets for lanthanide atoms with applications to the Ce diatom and LuF_3. J Phys Chem A 2008;112:11431–11435.

[47] Kutzelnigg W. r12-Dependent terms in the wave function as closed sums of partial wave amplitudes for large l. Theor Chim Acta 1985;68:445–469.

[48] Hotop H, Lineberger WC. Binding energies in atomic negative ions. J Phys Chem Ref Data 1975;4:539.

[49] Feller D, Davidson ER. The electron affinity of oxygen: a systematic configuration interaction approach. J Chem Phys 1989;90:1024–1030.

[50] Kaufmann K, Baumeister W, Jungen M. Universal Gaussian basis sets for an optimum representation of Rydberg and continuum wave functions. J Phys B: At Mol Opt Phys 1989;22:2223–2240.

[51] Mastalerz R, Widmark PO, Roos BO, Lindh R, Reiher M. Basis set representation of the electron density at an atomic nucleus. J Chem Phys 2010;133:Art. 144111.

7

SECOND QUANTIZATION AND MULTICONFIGURATIONAL WAVE FUNCTIONS

7.1 SECOND QUANTIZATION

In multiconfigurational work, it is necessary to represent wave functions, and operators on wave functions, that are described by a large number of configuration functions. The basic tool is then known as *second quantization*. This refers to the use of linear combinations of functions that belong to a special type of Hilbert space, namely, the *Fock space*. This space consists of *all* wave functions with $0, 1, 2, \ldots$ particles that can be formed from a given set of one-electron basis functions. For our purpose, the particles are electrons. The basis set can be infinite, in formal work, but is obviously finite in calculations. The unique wave function with 0 electrons represents the vacuum state ($|0\rangle$ or $|vac\rangle$), and from this, all other states are defined by using creation operators.

Assume that a basis set of molecular orbitals $\psi_1, \psi_2, \ldots, \psi_N$ is given. They may represent physically meaningful one-electron states, but they do not have to—they could be just computationally convenient basis functions. From these orbitals, a set of $(2^N - 1)$ *Slater determinant (SD) functions* can be formed, for example, for three electrons,

$$|ijk\rangle \overset{\text{def}}{=} \frac{1}{\sqrt{3!}} \begin{vmatrix} \psi_i(x_1) & \psi_j(x_1) & \psi_k(x_1) \\ \psi_i(x_2) & \psi_j(x_2) & \psi_k(x_2) \\ \psi_i(x_3) & \psi_j(x_3) & \psi_k(x_3) \end{vmatrix}.$$

Multiconfigurational Quantum Chemistry, First Edition.
Björn O. Roos, Roland Lindh, Per Åke Malmqvist, Valera Veryazov, and Per-Olof Widmark.
© 2016 John Wiley & Sons, Inc. Published 2016 by John Wiley & Sons, Inc.

By adding to this set the formal vacuum state, we obtain the basis for the (2^N)-dimensional Fock space. Even if we normally do not consider states with varying number of electrons, it turns out that also the number-preserving operators we use in quantum chemistry can be conveniently handled in terms of only the set of *creation* operators $\hat{a}_i^\dagger$, and their Hermitian conjugates, the *annihilation* operators $\hat{a}_i$. Moreover, the many-particle basis functions, usually SDs or spin-coupled *configuration functions*, CFs, will then be represented in a way that is suitable for machine computations. The creation operators are defined as follows. Let X be a symbol that is either empty or a sequence of arbitrary orbital indices. Then

$$\hat{a}_i^\dagger |vac\rangle = |i\rangle, \quad \text{and} \quad \hat{a}_i^\dagger |X\rangle = |iX\rangle.$$

Thus, the creation operator acts on any n-particle SD to produce an $n + 1$-particle SD.

Example:

$$\hat{a}_i^\dagger |jk\rangle = \hat{a}_i^\dagger \frac{1}{\sqrt{2!}} \begin{vmatrix} \psi_j(x_1) & \psi_k(x_1) \\ \psi_j(x_2) & \psi_k(x_2) \end{vmatrix}$$

$$= \frac{1}{\sqrt{3!}} \begin{vmatrix} \psi_i(x_1) & \psi_j(x_1) & \psi_k(x_1) \\ \psi_i(x_2) & \psi_j(x_2) & \psi_k(x_2) \\ \psi_i(x_3) & \psi_j(x_3) & \psi_k(x_3) \end{vmatrix} = |ijk\rangle.$$

The following properties of a determinant are well known:

1. Swapping any two columns, or any two rows, the determinant will change sign.
2. The determinant is linear in the elements of any specified row, and linear in the elements of any specified column.
3. The determinant of a diagonal matrix is equal to the product of the diagonal elements.

One can conclude that the Slater determinant, regarded as a wave function, will change sign when any two particle indices are swapped, so that it will describe antisymmetric particles; that this antisymmetry also holds for interchange of any orbital indices in the SD, so that the SD is 0 if any two orbital indices are the same. The set of independent SDs is thus those that have any number of distinct orbital indices in some predetermined order. There will be $\binom{n}{k}$ of these if there are n orbitals and k electrons. This binomial coefficient is $\frac{n!}{k!(n-k)!}$ which explains the very rapid increase when n is large. These SDs are orthonormal if the orbitals are.

7.2 SECOND QUANTIZATION OPERATORS

One of the points in using the second quantization is that the operators will not depend on any "irrelevant" parts or properties of the wave functions.

For example, the dipole moment operator (in the x direction) is defined in the "first quantization" (FQ) by the requirement that

$$\hat{\mu}_x^{FQ}\Psi(\mathbf{r}_1, \ldots, \mathbf{r}_N) = \sum_{n=1}^{N} x_n \Psi(\mathbf{r}_1, \ldots, \mathbf{r}_N)$$

for any N-electron wave function, where $N = 1, 2, \ldots, N$. This is the "first quantization" form of the operator. Assuming an orthonormal orbital basis, consider the set of all SDs constructed from this basis—this set is then an orthonormal basis for the space of all linear combinations of such SDs. Let's call this the "Full CI" space for short. Also, the set of linear operators on this space is itself a linear (vector) space. Let's call that the "Full CI operator space."

We can define the same operator in second quantized form (SQ) as

$$\hat{\mu}_x^{SQ} = \sum_{pq} \langle \varphi_p | x | \varphi_q \rangle \hat{a}_p^\dagger \hat{a}_q. \tag{7.1}$$

This operator is exactly equivalent to the first one, provided that the orbital basis is complete. Also, if the orbital basis is *not* complete, the results of applying the two operators above can differ only by some function orthonormal to the SD space. Thus, the second quantized form is actually identical to the first quantization operator *projected* on the Full CI operator space.

Less transparent, but very useful, operators are the "electron field operators"

$$\hat{\psi}(\mathbf{r}) = \sum_p \psi_p^*(\mathbf{r})\hat{a}_p \quad \text{and} \quad \hat{\psi}^\dagger(\mathbf{r}) = \sum_p \psi_p(\mathbf{r})\hat{a}_p^\dagger \tag{7.2}$$

that can be regarded as annihilating or creating an electron at position $\mathbf{r}$. This may be regarded as strange; there can be no operators that create or annihilate an electron in a single point. But it is readily seen that the definition above is in fact defining proper operators for any finite set of orbitals, and the resulting formulae will then work also in the limit of a complete space. Using the field operators, the expression of a one-electron matrix element becomes

$$\left\langle \Psi_1 \middle| \int \hat{\psi}^\dagger(\mathbf{r})v(\mathbf{r})\hat{\psi}(\mathbf{r}) \, d^3\mathbf{r} \middle| \Psi_2 \right\rangle$$

$$= \left\langle \Psi_1 \middle| \int \sum_p \psi_p(\mathbf{r})\hat{a}_p^\dagger v(\mathbf{r}) \sum_q \psi_q^*(\mathbf{r})\hat{a}_q \, d^3\mathbf{r} \middle| \Psi_2 \right\rangle$$

$$= \sum_{pq} \int \psi_p(\mathbf{r})v(\mathbf{r})\psi_q^*(\mathbf{r}) \, d^3\mathbf{r} \langle \Psi_1 | \hat{a}_p^\dagger \hat{a}_q | \Psi_2 \rangle$$

$$= \sum_{pq} \langle \psi_p | v | \psi_q \rangle \langle \Psi_1 | \hat{a}_p^\dagger \hat{a}_q | \Psi_2 \rangle$$

verifying the earlier statement, Eq. 7.1.

Similarly, for the two-electron repulsion matrix element is,

$$\left\langle \phi_1 \left| \int \int \frac{1}{|\mathbf{r} - \mathbf{r}'|} \hat{\psi}^\dagger(\mathbf{r}')\hat{\psi}^\dagger(\mathbf{r})\hat{\psi}(\mathbf{r})\hat{\psi}(\mathbf{r}') \, d\mathbf{r}^3 \, d\mathbf{r}'^3 \right| \phi_2 \right\rangle.$$

The definition of the Hamiltonian in second quantization, using creation and annihilation operators, is then

$$\hat{H} = \sum_{pq} \langle \psi_p|\hat{h}|\psi_q \rangle \, \hat{a}_p^\dagger \hat{a}_q + \frac{1}{2} \sum_{pqrs} \langle pr|qs \rangle \hat{a}_p^\dagger \hat{a}_r^\dagger \hat{a}_s \hat{a}_q,$$

where $\hat{h}$ is the one-electron part of the Hamiltonian, containing also the kinetic energy part, in a similar way as the one-electron potential energy. There is thus a direct link, via field operators, from first-quantization expressions involving directly the coordinates of particles, to second-quantization operators, and bra-ket formalism.

Using creators and annihilators, the equations can be evaluated expressing Slater determinants as a string of creators acting on the vacuum determinant. Basically, creators are "pushed to the left" and annihilators are "pushed to the right," and wherever a creator and an annihilator interchange their places, a Kronecker delta is generated, using the formula

$$\hat{a}_q \hat{a}_p^\dagger = \delta_{pq} - \hat{a}_p^\dagger \hat{a}_q$$

until an annihilator is next to the vacuum state on the right-hand side (or a creator on the left-hand side) that produces zero. In effect, the creators and annihilators are vanishing, leaving behind Kronecker deltas.

Example: The matrix element of the one-electron Hamiltonian evaluated for two SDs, which differ in one orbital:

$$\langle 32|\hat{h}|12 \rangle = \sum_{pq} \langle \phi_p|\hat{h}|\phi_q \rangle \left\langle vac \left| \hat{a}_3 \hat{a}_2 \hat{a}_p^\dagger \hat{a}_q \hat{a}_2^\dagger \hat{a}_1^\dagger \right| vac \right\rangle$$

$$= \sum_{pq} h_{pq} \left\langle vac \left| (-\delta_{p3}\hat{a}_2 + \delta_{p2}\hat{a}_3)(-\delta_{1q}\hat{a}_2^\dagger + \delta_{2q}\hat{a}_1^\dagger) \right| vac \right\rangle$$

$$= h_{31}. \tag{7.3}$$

A couple of steps were bypassed in this derivation, which is intended to show the principles. In practice, numerous shortcuts are used.

During the optimization of a wave function, represented by CI coefficients or related parameters (e.g., Coupled Cluster amplitudes), terms in the Hamiltonian are repeatedly acting on the wave function. Thus, it is natural to use the same orbitals also in second quantization formulas used to compute density matrices or properties afterward. An exception is, for example, when computing matrix elements over a set of wave functions using different orbitals—then other orbital sets may be used to advantage, see Chapter 10. Thus, it should be noted that the choice of orbitals is not always carved in stone or remain fixed, once used in some particular calculation.

7.3 SPIN AND SPIN-FREE FORMALISMS

The equation for the Hamiltonian above is valid for an orthogonal basis of general spin-orbitals. The one-electron matrix elements (MO-integrals) in the integral are

$$\left\langle \psi_p | \hat{h} | \psi_q \right\rangle = \sum_{\sigma \in \{\alpha, \beta\}} \int_{\mathbf{r} \in R^3} \frac{1}{2} \nabla \psi_p^*(\mathbf{r}, \sigma) \cdot \nabla \psi_q(\mathbf{r}, \sigma) + \psi_p^*(\mathbf{r}, \sigma) v(\mathbf{r}) \psi_q(\mathbf{r}, \sigma) \, d\mathbf{r}^3.$$

(7.4)

Here, we have assumed that the orbital functions depend on spin as well as on position, and that the spin is given as a variable σ that takes on either of two values, α or β. Other common conventions are, for example, to use α and β as names of two spin functions, which depend on a formal spin variable taking the two values $m_s = \pm\frac{1}{2}$, or to use orbitals that are 2×1 matrices with two elements that are ordinary orbital functions; the upper for $m_s = \frac{1}{2}$ and the lower for $m_s = -\frac{1}{2}$ (so-called two-component wave functions).

For simplicity, let us specialize to use the same set of spatial orbitals for α as for β spin, and to agree that in expressions such as that above, letting the symbol x denote the composite variable $(\mathbf{r}, \sigma)$, the integral symbol stand for also a sum over spin if the integration measure is written as dx, and to use merely $\psi_p(\mathbf{r})$ in spin-independent contexts. Moreover, we will often use simply ψ_p, or even just p, in bra and ket symbols. With these conventions, we may write matrix elements more concisely as

$$\left\langle \psi_p | \hat{h} | \psi_q \right\rangle = \int \frac{1}{2} \nabla \psi_p^*(\mathbf{r}) \, \nabla \psi_q(\mathbf{r}) + \psi_p^*(\mathbf{r}) v(\mathbf{r}) \psi_q(\mathbf{r}) \, d\mathbf{r}^3,$$

(7.5)

$$\langle pr|qs \rangle = \int \int \psi_p^*(\mathbf{r}_1) \psi_r^*(\mathbf{r}_2) \frac{1}{r_{12}} \psi_r(\mathbf{r}_1) \psi_s(\mathbf{r}_2),$$

(7.6)

where r_{12} is the distance $|\mathbf{r}_1 - \mathbf{r}_2|$. The Hamiltonian is now

$$\hat{H} = \sum_{pq} \left\langle \psi_p | \hat{h} | \psi_q \right\rangle \sum_{\sigma} \left(\hat{a}_{p\sigma}^\dagger \hat{a}_{q\sigma} \right) + \frac{1}{2} \sum_{pqrs} \langle pr|qs \rangle \sum_{\sigma, \sigma'} \left(\hat{a}_{p\sigma}^\dagger \hat{a}_{r\sigma'}^\dagger \hat{a}_{s\sigma'} \hat{a}_{q\sigma} \right).$$

In the two-electron expressions, you must pay attention to the order of indices. Other common symbols are

$$(pq, rs) \equiv \langle pr|qs \rangle$$

and

$$\hat{E}_{pq} = \sum_{\sigma} \left(\hat{a}_{p\sigma}^\dagger \hat{a}_{q\sigma} \right)$$

(7.7)

$$\hat{E}_{pq,rs} \equiv \hat{E}_{pq} \hat{E}_{rs} - \delta_{rq} \hat{E}_{ps} = \sum_{\sigma, \sigma'} \left(\hat{a}_{p\sigma}^\dagger \hat{a}_{r\sigma'}^\dagger \hat{a}_{s\sigma'} \hat{a}_{q\sigma} \right)$$

(7.8)

and finally, the two-electron operator in the Hamiltonian is

$$\frac{1}{2} \sum_{pqrs} (pq, rs) \; \hat{E}_{pq,rs}.$$

The spin is now hidden away; the effects of spin on the permutation properties are taken care of in the summations, by evaluation rules for the matrix elements over spin-adapted CSFs of the "excitation operators," $\hat{E}_{pq}$ and $\hat{E}_{pq,rs}$. (If one really wants to use spin-dependent operators, one must resort to using the primitive creation and annihilation operators).

The GUGA (Graphical Unitary Group Approach) [1, 2] is one of the two most common schemes for using spin-adapted CSFs instead of individual Slater Determinants (The other is the so-called Symmetric Group approach). The "graphical" characterization is because there is a well-known, very useful, depiction of the Full CI space in the form of a graph.

The (spatial) orbitals are ordered, and the graph consists of nodes arranged at different levels, starting with a single node at level 0. Each node corresponds to a subspace of the Full CI space. At level 1, the nodes correspond to the subspaces where only orbital 1 can be occupied, and each node corresponds to a specific number n of electrons and total spin S: At level 1, we have the possibilities $(n, S) = (0, 0)$, $(1, 1/2)$, or $(2, 0)$. At each level, the possible nodes can be reached only from some of the lower nodes: these possibilities form the *arcs* or edges of the graph. A CSF space with specified number of electrons and specified spin, and using a specified set of orbitals, corresponds to just one node; every CSF in that space corresponds to a specific sequence of arcs through the graph, connecting the vacuum level at level 0 with this node.

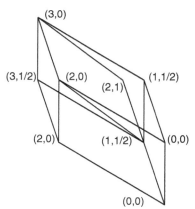

7.4   FURTHER READING

• E.R. Davidson, "Reduced Density Matrices in Quantum Chemistry", Academic Press, New York, 1976.

- T. Helgaker, P. Jørgensen, and J. Olsen, "Molecular Electronic-Structure Theory", John Wiley & Sons, Ltd, Chichester, 2000.
- B.P. Rynne and M.A. Youngson, "Linear Functional Analysis", Springer-Verlag, London, 2007.

7.5 REFERENCES

[1] Paldus J. Group theoretical approach to the configuration interaction and perturbation theory calculations of atomic and molecular systems. J Chem Phys 1974;61:5321–5330.
[2] Shavitt I. The graphical unitary group approach and its application to direct configuration interaction calculations. In: Hinze J, editor. The Unitary Group for the Evaluation of Electronic Energy Matrix Elements. Berlin: Springer-Verlag; 1981. p. 51–99.

8

ELECTRON CORRELATION

Correlation of the motion or distribution of the smallest constituents of a larger system results in effects that go beyond what can be understood from a strict mean-field picture. An understanding of such phenomenon is necessary to explain, for example, how it is possible that dispersed particles in a solution can have equal charges but yet attract each other. Most chemists are familiar with such effects, since thermodynamics and statistical physics play a major role in chemistry. These effects are generally studied by detailed statistical analysis of, for example, molecular dynamic simulations.

Similarly, regarding the electrons in a molecular system as constituting a fixed charge cloud would be misleading, since that charge density is merely an expectation number, while the very strong electrostatic repulsion depends on a joint probability.

The special case of electrons in a molecular system presents its own issues, different from that in general statistical mechanics, and different from that used in general statistics and probability theory. Like in the classical case, correlation can be analyzed by computing probability distributions, but in contrast to the classical case, so-called reduced density matrices are the most versatile tools for describing correlation in quantum mechanical systems.

8.1 DYNAMICAL AND NONDYNAMICAL CORRELATION

Electron correlation is customarily divided into dynamical and nondynamical, but there is no strict definition of these terms. In the context of quantum chemistry

Multiconfigurational Quantum Chemistry, First Edition.
Björn O. Roos, Roland Lindh, Per Åke Malmqvist, Valera Veryazov, and Per-Olof Widmark.
© 2016 John Wiley & Sons, Inc. Published 2016 by John Wiley & Sons, Inc.

calculations, these terms are mainly used to describe the different ability of different methods to capture significant correlation effects. To take account of nondynamical correlation then means that energetically close or degenerate electronic state are given an evenhanded description by using several (or many) configurations, valence-bond structures, or Slater determinants to describe them; the simplest example would be for an atomic or diatomic radical, where the multiplet structure requires several determinants for its description. This can be done already by SCF procedures, if a predetermined electronic structure with more than one determinant is optimized by only adjusting the orbitals, as is done by so-called open-shell Hartree–Fock that has been extensively used in atomic physics. But in general, a similar situation is at hand when such electronic structure occur, for example, toward dissociation, "accidental" near degeneracy, open d-shells, reactions that are Hartree–Fock forbidden or whose description by single-orbital methods require symmetry breaking, or cause multiple local minima in the orbital optimization when single-determinant methods are attempted. In photochemistry, such effects are rule rather than exception.

By contrast, the dynamical correlation describes a situation where double (or higher) excitations from strongly occupied shells to weakly occupied correlating orbitals can adequately describe the stabilizing effect of allowing electrons to avoid coming too close, when the orbital density in a mean field picture would allow such close encounters. Dynamical correlation can also be described without correlating orbitals, for example, adding a correlation potential to the mean field (bare Coulomb or including exchange) using a density functional, or in some highly accurate methods where the wave functions contain terms that depend directly on the interelectronic distances.

8.2   THE INTERELECTRON CUSP

When two electrons are close, the exact electronic wave function has an "interelectron cusp," a singularity in the derivative of the wave function, and of the two-electron density function, when two electrons coalesce. For $\mathbf{r}_1 \approx \mathbf{r}_2$, we may define the new position vectors $\mathbf{R}$ and $\mathbf{r}_{12}$ with $\mathbf{r}_1 = \mathbf{R} + \mathbf{r}_{12}/2$ and $\mathbf{r}_2 = \mathbf{R} - \mathbf{r}_{12}/2$. For fixed vectors $\mathbf{R}, \mathbf{r}_3, \cdots$, the wave function $\Phi(\mathbf{r}_1, \mathbf{r}_2, \mathbf{r}_3, \cdots, \mathbf{r}_n)$ is then a function of $\mathbf{r}_{12}$ only, which can be expanded using spherical harmonics with some leading quantum number l:

$$\Phi(\mathbf{r}_1, \mathbf{r}_2, \mathbf{r}_3, \cdots, \mathbf{r}_n) \sim r_{12}^l \left(1 + \frac{r_{12}}{(2l+2)} + \cdots \right).$$

Here, Φ is the spatial part of a wave function—the full wave function will in general be a sum of four such functions multiplied by spin functions, one of which is a singlet, and three are components of a triplet function. The spatial functions have l even, for a singlet spin function, and odd for a triplet spin function. The cusp is introduced by the $r_{12}/(2l+2)$ term. In general position, $l = 0(1)$ for singlet (triplet) pairs, but it can be higher (in increments of two). The asymptotic formula is not valid in case of three-body coalescence. The cusp is often called "the Kato cusp" [1], but the detailed description is by Pack and Brown [2], and it was of course known by those who actually performed explicitly correlated calculations, for example, Hylleraas already in 1928 [3].

Conventional quantum chemistry programs use one-electron orbitals to describe correlation, and this gives a poor description of the cusp region. There is no cusp as such, but also the surrounding "correlation hole" is only roughly reproduced. Kutzelnigg [4] showed that inclusion of a single term in the CI expansion, if this term was able to provide the cusp, gave a convergence of energy with asymptotic error $(L + \frac{1}{2})^{-7}$, rather than $(L + \frac{1}{2})^{-3}$ that is obtained by a conventional CI, if L denotes the highest L quantum number of the atomic basis, while the radial basis is saturated.

So-called "R12" and "F12" methods contain a correlation factor designed to give the correct interelectron cusp. These are generally complicated, but a large amount of work and ingenuity has made these methods available for calculations of chemical accuracy on systems as large as, for example, thiophene. For a review, see Ref. [5]. Such model functions are also used in Quantum Monte Carlo calculations and are then usually called "Jastrow factors" [6].

Thus, for largest accuracy, and in particular for calculating quantities that depend crucially on the pointwise accuracy of the two-electron density function at coalescence, "F12" type calculations are the natural choice. Nevertheless, CI expansions are also able to give a large part of the correlation energy. Most importantly, the missing part should not depend strongly on, for example, molecular structure.

Figure 8.1 shows the effect of the Kato cusp on a system that allows simple calculations. This is a two-electron system but not an atom; the electrons are held confined by a harmonic force field.

For this system, the center of mass position is statistically independent of the interelectron vector $\mathbf{r}_{12}$, and for the ground state the dependence on that vector is only through the distance r_{12}. The independent variable is thus the distance r_{12}, and the figure shows that close to coalescence, there is indeed a cusp where

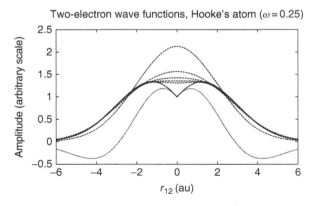

Figure 8.1 The two-electron wave function of Hooke's 'atom' (see text).

$\psi(r_{12}) \approx \psi(0)(1 + r_{12}/2)$ (the solid line is the singlet ground state). Also shown (dashed curves) are a number of CI approximations, which give an imperfect reduction in the two-electron density. An additional curve (dotted) for an excited state shows that the cusp is similar also for that state. These observations are general but would be difficult to see in corresponding calculations for a genuine two-electron atom or ion.

The correlation due to the interelectronic cusp is called "dynamical correlation," and other types of correlation are then "nondynamical." Figure 8.2 shows in another way the two-electron density in a valence region of Be. Outside of the central $(1s)^2$ core, one of the valence electrons is placed at a fixed position, as indicated in the figure, which is a contour plot of the two-electron density as function of the position of the other electron.

It is clearly seen how one electron is kept from approaching the other by a large and wide depression of the two-electron density. The effect is obtained in a CI calculation

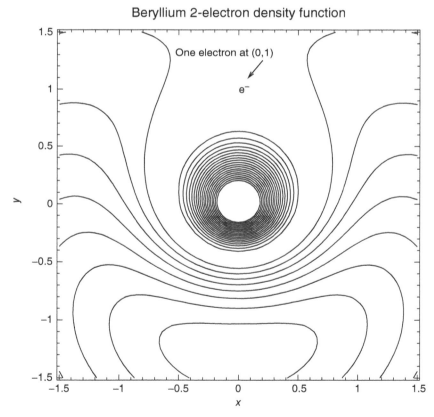

Figure 8.2 The two-electron density function for the valence electrons of the Be atom (see text).

with just a few configurations (the $(1s^2 2s^2)^1 S$ and $(1s^2 2p^2)^1 S$), and it is so large that the interelectron cusp is quite insignificant. This is the hallmark of a nondynamical correlation. While the cusp would give the necessary accuracy for, for example, spectroscopic use, it would need to be based on a wave function where the nondynamical correlation is already present.

The "correlation energy" is a quantity that depends, for its definition, on what is considered to be the independent particle model. After Löwdin, this is conventionally taken to be the Hartree–Fock model, but that is not always the most relevant choice; if not, it is best to specify, and to avoid unqualified use of the term "correlation energy." The correlation, in the statistical sense, of electrons due to the antisymmetry of the wave function is occasionally called "Pauli correlation" or "Fermi correlation" but is not considered as correlation when the independent particle model is Hartree–Fock. However, consider a simple open-shell atom with a multiplet splitting of a configuration. When treated by open-shell Hartree–Fock, there is no fundamental difference between the way that the exchange terms of a closed-shell system is treated, versus the electrostatic splitting of the configuration manifold into electronic terms; the exchange is replaced by similar integrals that enter with predetermined coefficients dictated by the assumed symmetry properties of the electronic wave function. Yet the energy gained, when compared to a more or less irrelevant single-determinant wave function, is by convention called the correlation energy (provided that the comparison is done for the lowest state). The correlation effects that are described by using the predetermined linear combination of Slater determinants for the individual states of the multiplet are termed nondynamical correlation, but there may be no meaningful correlation energy associated with each state. Also, each set of degenerate states may be decomposed in infinitely many ways into orthonormal, noninteracting states; sometimes, they can be chosen such that a few of them can be single-determinant states, but this is not a fundamental distinction; the states in a degenerate set are equivalent by symmetry.

In this extreme case, the need for multiconfigurational (or at least multideterminant) treatment is a consequence of symmetry. But the same applies also when the symmetry is broken, or there is no obvious symmetry, whenever electronic states can become nearly degenerate. This is usually the case when electronically excited states are studied, as in photochemistry. But it also happens when bonds are formed or broken.

8.3 BROKEN BONDS. $(\sigma)^2 \rightarrow (\sigma^*)^2$

The following figures need a moment of attention—it is a contour plot of a two-electron wave function, as a function of *two* electron positions. The figure shows contour plots of the H_2 wave function for the special case that the electrons are both at the molecular axis, as a function of the two z-coordinates. The graphs are accurate; the isocurves are almost squares, due to the e^{-r} character of the orbitals close to the nuclei. In the first pair of figures, side by side (Figure 8.3), is the Hartree–Fock wave function and an accurate CI wave function, at the equilibrium

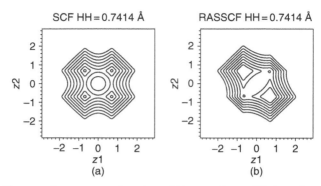

Figure 8.3 The two-electron wave function of H_2 near the equilibrium distance, from HF (a) and Full CI (b).

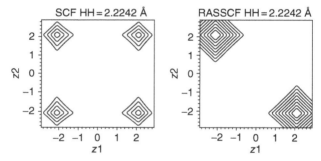

Figure 8.4 The same, but with the molecule stretched to about three times the equilibrium distance.

distance (HH $= 0.74\,\text{Å}$). In the second pair (Figure 8.4), the distance has been increased to $2.22\,\text{Å}$.

It is clear what happens in the Hartree–Fock calculation. The ground state wave function is $\Psi(\mathbf{r}_1, \mathbf{r}_2) = \sigma_g(\mathbf{r}_1)\sigma_g(\mathbf{r}_2)$ that has large values when the two factors $\sigma_g(\mathbf{r}_1)$ and $\sigma_g(\mathbf{r}_2)$ are large, which is when both electrons are close to either of the nuclei, but it does not matter if they are both close to the same nucleus, or if they are each close to a different nucleus. This does not matter much at the equilibrium distance. But when the bond is broken, it is definitely wrong. In that case, we expect a very low probability for the electrons to stay close to the same nucleus, and this is also clearly shown for the accurate wave function. In this case, that is a CI type wave function, and the low amplitude for the electrons to be close to the same nucleus is caused by a destructive interference of two configurations,

$$\Psi(\mathbf{r}_1, \mathbf{r}_2) \approx \frac{1}{\sqrt{2}}(\sigma_g(\mathbf{r}_1)\sigma_g(\mathbf{r}_2) - \sigma_u(\mathbf{r}_1)\sigma_u(\mathbf{r}_2)).$$

A moment's reflection shows that something similar may happen whenever a bond is stretched in similar situations. If a bond is lengthened, an accurate description of the electronic structure requires some amplitude for the configuration with a doubly occupied bonding orbital replaced by an antibonding orbital.

8.4 MULTIPLE BONDS, AROMATIC RINGS

When multiple bonds are broken, much the same happens, except that the bond-breaking process requires many more configurations for a qualitatively correct description. However, they can usually be regarded as a product of single-bond factors. In fact, well-characterized bonding orbitals are usually naturally associated with antibonding partners, and usually there is a remarkably precise relation between the natural occupation numbers of these orbitals. An often quoted example is the Cr_2 calculation by Roos [7]; see Figure 8.5.

To the left is seen the active natural orbitals of the ground state near the equilibrium structure. From the occupation numbers shown in the figure, it is a striking fact that the natural orbitals appear to occur in pairs, with occupation numbers adding up very closely to 2. Each such pair represents a correlated bond, consisting of one strongly occupied orbital, and one correlating orbital.

As shown to the right, this remains true for a large range of distances, until the bond is completely broken and we have 12 degenerate orbitals. At dissociation, each atom (considered separately) is described by an ensemble of 35 states, belonging to the $d^5s(^7S)$ multiplet. These states still form a "tangled" diatomic $^1\Sigma^+$ state, which requires several hundred configurations for proper description of the dissociation.

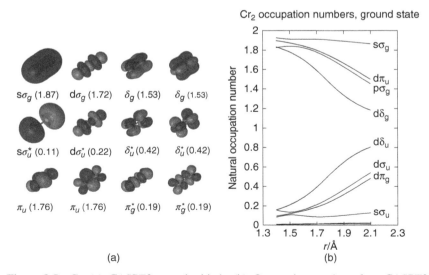

(a)

(b)

Figure 8.5 Cr_2 (a): CASPT2 natural orbitals; (b): Occupation numbers, from CASPT2.

This is a general observation for ordinary multiple bonds: a central σ bond contributes binding already at large distance. The π or δ bonds develop at shorter distance. The result is a compressed σ bond and a partially broken π or δ bond, at the equilibrium geometry. Such partially broken bonds cause multiconfigurational character of the ground state, and also is the reason for the photochemical activity of multiple bonds: the transitions from bonding to antibonding orbital are lower in energy and have large oscillator strength.

This description of nondynamical correlation in terms of bonding–antibonding pairs is true, however, only for the simplest wave function; it is called a *perfect pairing* wave function. Actually, even when occupation numbers add up in this striking manner, an accurate wave function contains many terms that are not just products of pairs. Moreover, excited and radical wave functions can be much more complicated. Nevertheless, it is true that wave functions with a large number of active orbitals can in principle be described by far less parameters than are needed for, for example, a CASSCF wave function, but it is not easy to find general or automated ways of using this fact. There is of course also dynamical correlation. The difference is not clear-cut, except perhaps operationally: The dynamical correlation is that part that is not obtained by a CASSCF wave function with a minimal active space. This distinction is not very useful. The dynamical correlation is conventionally described by the Kato cusp. Since that is really a point-like property, this is not very useful either. In conventional (not R12) calculations, this is characterized by needing a very large number of orbitals, if it is to be correctly described.

8.5 OTHER CORRELATION ISSUES

There is an additional effect, which is not so often considered. Dispersion interaction is usually considered mainly if weak intermolecular interactions are important. They are always present, and they are quite strong at shorter distance. However, they are automatically taken into account when including dynamical correlation in the sense of two-electron excitation. It is a matter of taste if one wishes to use the term dynamical correlation or not. It is similar to the effect of forming a cusp and its correlation hole, but instead of the electrons avoiding each other in a pairwise fashion, it is a composite effect of any regions of the electron cloud to show a dynamical polarization in response to the fluctuating field from another region. The total effect is the sum of a very large number of individual contributions, just as for the more conventional near-encounter (cusp-type) correlation.

From a large body of computational experience, one can distinguish a cusp contribution to the correlation energy, which is roughly 1 eV, or 20 kcal/mol, for each strongly occupied electron pair. Evidently, the core region then gives large contributions to the total correlation energy; yet this part is almost always neglected (except when the core structure is disturbed in the processes one wants to describe as in X-ray spectroscopy, EXAFS, or photoelectron spectroscopy).

However, there are good reasons for not correlating the core: its contribution to the dynamical correlation energy is large but almost constant across chemical processes. Since the wave function is optimized by minimizing the energy, a conventional quantum chemistry code would use all the basis functions that happen to overlap with the core region, including basis function tails from other atoms, in an attempt to optimize this region of space. A small fraction of the large correlation energy associated with the core region will be obtained, and this fraction will vary with geometry. Therefore, a large *basis set superposition error* (BSSE), swamps the accuracy to be gained, unless large and specialized basis sets are used. For high accuracy, one can still profitably include part of the core correlation: excitations that remove only one core electron are better behaved and may be regarded as a dispersion interaction between core and valence.

The distinction between dynamical and nondynamical correlation can be illustrated, for example, by the following simple calculations on the CO molecule (Figure 8.6).

SCF includes no correlation at all. In this case, the dissociation is 'allowed' (closed shell to closed shell), but it *should* be to *radicals*. CASSCF includes nondynamical correlation only, which is enough for low accuracy and gives qualitatively correct dissociation to ^3P atoms. CASPT2 adds in the dynamical correlation, necessary for a quantitatively correct result: the dynamical correlation depends on the electronic structure of the root state, so it changes considerably with dissociation.

The two kinds of correlation are not independent. When nondynamical correlation is included in a calculation, the valence electrons may already avoid each other, and the Kato cusps get smaller. In excited states, the opposite can take place, and the dynamical correlation may increase. This is quite understandable if one thinks of a *charge transfer* state, where the atoms involved may have quite different opportunities for dynamical correlation compared to other states.

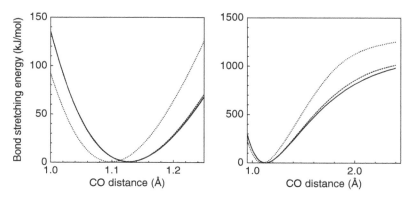

Figure 8.6 Potential curves of CO computed by SCF (dotted), CASSCF (dashed), and CASPT2 (solid).

8.6 FURTHER READING

- C.J. Cramer, "Essentials of Computational Chemistry", John Wiley & Sons, Ltd, Chichester, 2002.
- T. Helgaker, P. Jørgensen, and J. Olsen, "Molecular Electronic-Structure Theory", John Wiley & Sons, Chichester, 2000.
- A. Szabo and N.S. Ostlund, "Modern Quantum Chemistry", Dover Publication, Inc., Mineola (NY), 1996.

8.7 REFERENCES

[1] Kato T. On the eigenfunctions of many-particle systems in quantum mechanics. Commun Pure Appl Math 1957;10:151–177.

[2] Pack RT, Brown WB. Cusp conditions for molecular wavefunctions. J Chem Phys 1966;45:556–559.

[3] Hylleraas EA. Über den Grundzustand des Heliumatoms. Z Phys 1928;48:469–494.

[4] Kutzelnigg W. r12-dependent terms in the wave function as closed sums of partial wave amplitudes for large l. Theor Chim Acta 1985;68:445–469.

[5] Klopper W, Manby FR, Ten-No S, Valeev EF. R12 methods in explicitly correlated molecular electronic structure theory. Int Rev Phys Chem 2006;25:427–468.

[6] Jastrow R. Many-body problems with strong forces. Phys Rev 1955;98:1479–1483.

[7] Roos BO. The ground state potential for the chromium dimer revisited. Collect Czech Chem Commun 2003;68:265–274.

9

MULTICONFIGURATIONAL SCF THEORY

This chapter introduces the multiconfigurational SCF (MCSCF) theory and consists of four sections. The first section gives a background for the need to develop an MCSCF theory. Here are presented a number of chemical and molecular cases in which it is clear that a single-configurational approach is insufficient for a qualitative correct description. The second section is devoted to the technique of determining the MCSCF wave function. The section also discusses some aspects of different representations and higher order derivatives with respect to the wave function parameters. The third section presents two special types of MCSCF wave functions—the complete active space SCF (CASSCF) and restrictive active space SCF (RASSCF)—that have had tremendous impact on the use of the MCSCF method. Finally, we conclude this chapter with a discussion of some general rules to be applied in the selection of active orbitals for CASSCF and RASSCF calculations.

9.1 MULTICONFIGURATIONAL SCF THEORY

In this section, we discuss the multiconfigurational (MC) approach in terms of a partitioning of the exact wave function (in a finite one-particle basis), as represented by a linear combination of all possible electron configurations—the full Configuration Interaction (CI) wave function. The electronic configurations are here represented either by Slater determinants or by their spin-adapted counterpart, the so-called Configuration State Functions (CSF). The multiconfigurational approach is based on a

Multiconfigurational Quantum Chemistry, First Edition.
Björn O. Roos, Roland Lindh, Per Åke Malmqvist, Valera Veryazov, and Per-Olof Widmark.
© 2016 John Wiley & Sons, Inc. Published 2016 by John Wiley & Sons, Inc.

very simple concept: the partitioning of the wave function into a large and a small part:

$$\Psi = C_L\Psi_L + C_S\Psi_S, \tag{9.1}$$

where Ψ_L comprises the most important electronic configurations in the full CI expansion of the wave function, while the rest is included in Ψ_S. C_L should therefore be close to one. Almost all molecular wave functions can be partitioned in this way. Ψ_L is built from a limited number of molecular orbitals (the wave function for H_2 is a simple example with two orbitals). This limited set of molecular orbitals are popularly referred to as the *active space*. The subsequent MCSCF wave function is then generated as a full CI in the subspace of these active orbitals. Expansion coefficients and molecular orbital coefficient are established in accordance with the variational principle. The second part, Ψ_S, is determined in a subsequent step using configuration interaction techniques or perturbation theory. The partitioning of the wave function into a large and a small component is conceptually simple, but it is far from trivial in practice. We need to define the orbitals to be included in the MCSCF wave function and their basic structure (occupation). This requires knowledge about the general electronic structure of the system we are studying. We spend quite some time to discuss this problem and show how it may be solved in different types of applications. One must remember that the orbital space must be large enough to include all electronic situations/configurations that will occur for the energy surface(s) we are studying. If we were only studying the molecule in its ground state close to equilibrium, it could often suffice to use one CSF in Ψ_L, the Hartree–Fock configuration. However, normally we are interested in larger parts of the energy surface(s), for example, when we study a chemical reaction in the ground state and/or in excited states. So, we need an active space that can simultaneously describe more than one electronic configuration.

The first thing we do is to explore some simple cases, where MC wave functions are needed, and how they can be constructed. We then become more formal and set up the wave function by partitioning the orbital space into an inactive, an active, and a virtual part. We briefly discuss how we can determine the CI expansion coefficients and the molecular orbital. Finally, we discuss the construction of the most important type of MCSCF wave function that is used today, the Complete Active Space SCF function, and end with an extensive discussion of how to choose active orbitals for different chemical problems.

9.1.1 The H_2 Molecule

We start with a more detailed analysis of the H_2 molecule. It is surprising how much understanding of the electronic structure and different ways to approximate the wave function one can obtain by studying this little molecule. We use a minimal basis set in this study with one $1s$ function on each atom: $1s_A$ and $1s_B$ on the atoms A and B, respectively. From these atomic functions, we can construct two spatial molecular orbitals:

$$\varphi_j = N_j(1s_A \pm 1s_B), \tag{9.2}$$

where $j = 1, 2$, representing the symmetric and antisymmetric combinations. The atomic functions fulfill the symmetry property:

$$\hat{\imath} 1s_A = 1s_B. \tag{9.3}$$

The two molecular orbitals can therefore be labeled as symmetric (gerade) or antisymmetric (ungerade) with respect to the inversion operator, $\hat{\imath}$. They are also symmetric with respect to rotation around the molecular axis and can be labeled σ. The two MOs can thus be written:

$$\sigma_g = N_g(1s_A + 1s_B),$$
$$\sigma_u = N_u(1s_A - 1s_B). \tag{9.4}$$

The bonding orbital is in the RHF model assumed to be doubly occupied leading to a total wave function of the form:

$$\Psi_1(r_1, r_2) = \sigma_g(r_1)\sigma_g(r_2)\Theta_{2,0}, \tag{9.5}$$

which we shall simply denote as $(\sigma_g)^2$. $\Theta_{2,0}$ is the singlet $(S = 0)$ antisymmetric spin function for two electrons:

$$\Theta_{2,0} = \sqrt{\frac{1}{2}}(\alpha_1\beta_2 - \beta_1\alpha_2). \tag{9.6}$$

The exact ground state wave function for H_2 is at the equilibrium geometry strongly dominated by a single electronic configuration, $(\sigma_g)^2$. The σ_g orbital is a bonding orbital with an increased electron density in the region between the two atoms. The double occupation of this orbital corresponds to the Restricted Hartree–Fock (RHF) approximation for the H_2 molecule. The RHF model leads to a reasonably accurate description of H_2 around the equilibrium geometry: computed bond distance is 0.735 Å (Experiment 0.746 Å) and the bond energy is 84 kcal/mol (Experiment 109 kcal/mol). It is typical for the HF model that it is able to describe closed-shell systems around their equilibrium geometry rather well. The correlation energy is only a small fraction of the total energy, but it is strongly distance dependent, which explains the error in the computed bond energy (there is no correlation energy at all for two separated hydrogen atoms).

The bond energy given above has been obtained by subtracting from the RHF energy for H_2 the energy of two separated hydrogen atoms (-627.5 kcal/mol). Suppose instead that we would use the RHF model to compute the potential curve for the dissociation of H_2, then the first thing to note is that the form of the molecular orbitals (see Eq. 9.4) are independent of the internuclear distance. The same form of the wave function (see Eq. 9.5) is thus obtained also for the separated atoms. Let us expand this wave function as products of the atomic orbitals $1s_A$ and $1s_B$:

$$\Psi_1(r_1, r_2) = N_1^2[1s_A(r_1)1s_A(r_2) + 1s_A(r_1)1s_B(r_2) +$$
$$1s_B(r_1)1s_A(r_2) + 1s_B(r_1)1s_B(r_2)]. \tag{9.7}$$

We note that this wave function contains the so-called ionic terms, contributions where both electrons are located at the same atom. These terms are clearly unphysical at large separations, since they correspond to the dissociation to $H^+ + H^-$, which has an energy around 320 kcal/mol above $H + H$. It is only the second and third term in the wave function above that describe correctly the dissociated homolytic products.

It is a typical feature of the RHF model to include these "ionic structures" in fixed proportions into the wave function. Consequently, the model cannot in general be used to describe dissociation processes, in particular not homolytic processes. The potential curve corresponding to this wave function, Ψ_1, will actually end up with an energy around 160 kcal/mol above the true energy at the limit of infinite separation.

Is there a remedy to this problem? Well, the most straightforward solution is to introduce coefficients in front of the two different terms in Ψ_1 (the symmetry of H_2 defaults Ψ_1 to two unique terms while for a general heteronuclear diatomic system there will of course be four unique terms and coefficients) and write

$$\Psi_{VB} = C_{Ion}\Psi_{Ion} + C_{Cov}\Psi_{Cov}, \tag{9.8}$$

where we have used the label VB to indicate that this is the valence bond formulation of the wave function for the hydrogen molecule. The ionic and covalent parts are given as

$$\Psi_{Ion}(r_1, r_2) = N_{Ion}[1s_A(r_1)1s_A(r_2) + 1s_B(r_1)1s_B(r_2)]\Theta_{2,0},$$
$$\Psi_{Cov}(r_1, r_2) = N_{Cov}[1s_A(r_1)1s_B(r_2) + 1s_B(r_1)1s_A(r_2)]\Theta_{2,0}. \tag{9.9}$$

The coefficients C_{Ion} and C_{Cov} in Eq. 9.8 can be varied to yield the correct wave function at the limit of infinite separation ($C_{Ion} = 0$ and $C_{Cov} = 1$). At equilibrium $C_{Ion} \approx C_{Cov}$, which reflects the fact that the RHF determinant dominates the wave function. The formulation in Eq. 9.8 forms the basis for the valence bond description of the chemical bond. In its description of the molecular wave function in terms of nonorthogonal atomic structures, it has an appealing chemical appearance. On the other hand, the formulation in terms of nonorthogonal (atomic) basis functions (orbitals) leads to a complicated mathematical structure, which so far has prevented large-scale applications in chemistry. Instead we shall look for a formulation in terms of orthogonal molecular orbitals. We introduce in addition to σ_g, the antibonding orbital σ_u. In terms of σ_g and σ_u, we can now write the wave function, Eq. 9.8, as

$$\Psi_{MC} = C_1\Psi_1 + C_2\Psi_2, \tag{9.10}$$

where Ψ_2 is the electronic configuration $(\sigma_u)^2$. This is the multiconfigurational (MC) molecular orbital formulation of the wave function for the chemical bond in H_2. It will correctly describe the entire potential surface (Figure 9.1). Close to equilibrium $C_1 \approx 1$ and $C_2 \approx 0$, while at large separations $C_1 \approx -C_2$. *The quantum chemical description of a chemical bond thus involves both the bonding and the antibonding orbital.* Another way of viewing the multiconfigurational wave function, Eq. 9.10, is

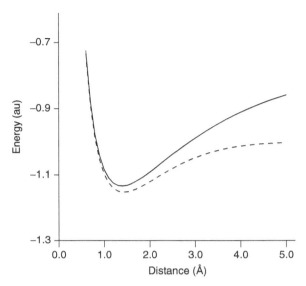

Figure 9.1 Potential curves for H_2 showing the erratic behavior of the RHF curve (solid line) as a function of the internuclear distance. For comparison, the MCSCF curve, which dissociates correctly, is also given (dashed line).

to note that the two configurations Ψ_1 and Ψ_2 are degenerate at infinite separation. Since the interaction between them is different from zero, strong mixing will occur with $C_1 = \pm C_2$. It is clear that the RHF model will not work in cases where more than one electronic configuration have the same, or nearly the same, energy. Below we show several examples of situations where near degeneracy occurs between different electronic configurations. A multiconfigurational treatment is then needed in order to obtain a qualitatively correct description of the electronic structure.

9.1.2 Multiple Bonds

The dissociation of a single bond can be described by a wave function composed of two electronic configurations. A more complex situation occurs in processes involving the simultaneous dissociation of several bonds. As an example, consider the triple bond in the nitrogen molecule N_2. The valence bond description at large separation arises from the coupling of two quartet nitrogen atoms (the $S = 3/2, M_S = 3/2$ electronic configuration of the nitrogen atom corresponds to 3 electrons in each p orbital) into an overall singlet. The corresponding MC wave function is obtained by a transformation from the $2p$ AO basis to the molecular orbitals (the z-axis along the bond axis):

$$3\sigma_g \propto 2p_{Az} - 2p_{Bz},$$

$$3\sigma_u \propto 2p_{Az} + 2p_{Bz},$$

$$3\pi_{ux} \propto 2p_{Ax} + 2p_{Bx},$$

$$3\pi_{gx} \propto 2p_{Ax} - 2p_{Bx},$$

$$3\pi_{uy} \propto 2p_{Ay} + 2p_{By},$$

$$3\pi_{gy} \propto 2p_{Ay} - 2p_{By}, \tag{9.11}$$

which we denote the "active orbitals." It is clear that the MC function will contain electronic configurations, which are up to sextuply excited with respect to the basic RHF configuration for N_2:

$$\Psi_{RHF} = (1\sigma_g)^2(1\sigma_u)^2(2\sigma_g)^2(2\sigma_u)^2(3\sigma_g)^2(1\pi_u)^4. \tag{9.12}$$

We shall not perform the somewhat elaborate calculation of the MC wave function in detail. Here we only note that the number of CSFs will increase very quickly with the number of "active" orbitals. In most cases, we do not have to worry about the exact construction of the MC wave function that leads to correct dissociation. We simply use all CSFs that can be constructed by distributing the electrons among the active orbitals. This is the idea behind the CASSCF method. The total number of such CSFs is for N_2 175 for a singlet wave function. A further reduction is obtained by imposing spatial symmetry. All these CSFs, of the CASSCF wave function, are not included in a wave function constructed from the valence bond structure for two ground state nitrogen atoms. A few terms are missing, since they correspond to different spin couplings—fragments while still coupled together to a singlet may not be in a quartet state and the fragments might not have a homogeneous distribution of the electrons (general valence bond theory could include these terms). Experience shows, however, that spin recouplings easily occur when a bond is formed, and it may therefore be dangerous to exclude these terms from the wave function, even if they are not needed for a correct description of the asymptotic wave function. The general conclusion is then that the MC wave function should comprise all the possible CSFs as is done in the CASSCF method.

9.1.3   Molecules with Competing Valence Structures

In the examples above, we considered the degeneracy effects that arises at the dissociation limit for a covalent chemical bond. Near degeneracy can also occur, for example, when one or more virtual orbitals have low energies. The beryllium atom was mentioned in the introduction as an example. Similar situations occur for some molecules in their ground state close to the equilibrium geometry. The RHF wave function for the ground state of a molecule can often be inferred from the chemical structure formula: each chemical bond and each lone pair corresponds to a doubly occupied (localized) orbital. The full closed-shell electronic configuration is obtained by adding orbitals for the core electrons. When the molecule possesses symmetry, it is usually straightforward to transform the occupied MOs to symmetry-adapted form.

However, there are systems for which it is not clear how to write down the structure formula. Such situations occur when all valences cannot be filled in a natural way. Several valence structures are then possible. It is not unexpected to find that the

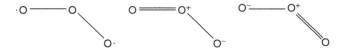

Figure 9.2 The competing valence structures for the ozone molecule.

RHF model then also runs into similar problems. Consider as an example the ozone molecule O_3. The most important valence structures for this molecule are represented in Figure 9.2.

The two last of these structures have an oxygen–oxygen double bond, at the expense of moving one electron to the end atom. The first structure corresponds to a diradical where two of the π electrons are unbound. The near degeneracy between these valence structures can in MO theory be traced to the near degeneracy between the two upper π molecular orbitals; π_2 (nonbonding) and π_3 (antibonding) (the molecule is labeled $O_B O_A O_C$):

$$\pi_1 = c_{11}\pi_A + c_{12}(\pi_B + \pi_C),$$
$$\pi_2 = c_{22}(\pi_B - \pi_C),$$
$$\pi_3 = c_{31}\pi_A + c_{32}(\pi_B + \pi_C). \tag{9.13}$$

The RHF electronic configuration is $(\pi_1)^2(\pi_2)^2$, but an MCSCF calculation reveals that the configuration $(\pi_1)^2(\pi_3)^2$ also has a large weight in the wave function:

$$\Psi = 0.89(\pi_1)^2(\pi_2)^2 - 0.45(\pi_1)^2(\pi_3)^2. \tag{9.14}$$

An analysis in terms of VB structures shows that this configurational mixing corresponds to approximately 40% biradical character in the wave function for ozone. The RHF wave function, on the other hand, contains only 12% of the biradical VB structure (the result was obtained using Hückel values for the coefficients of the orbitals 9.13). It is clear from these considerations that a correct treatment of the electronic structure for the ozone molecule must be based on a multiconfigurational wave function.

9.1.4 Transition States on Energy Surfaces

The calculation of energy surfaces for chemical reactions is an important challenge for quantum chemistry. Such surfaces often exhibit a saddle point somewhere between the minimum energy structure of the reactants and that of the products. The barrier is almost always due to a change of the dominating electronic configuration in the wave function, as shown in Figure 9.3. The transition state is an avoided crossing, a manifestation of at least two different electronic configurations being degenerate at the structure of the transition state.

Here E_I is the energy of the electronic configuration for the reactants (point R on the energy surface). This energy has a minimum around R and is repulsive outside

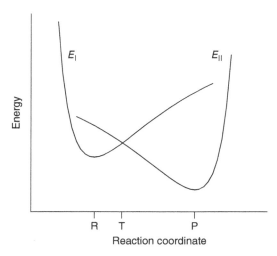

Figure 9.3   The potential energies for the reactant (E_I) and product (E_{II}) electronic configurations along the reaction coordinate.

this region. E_{II} is the corresponding energy curve for the electronic configuration of the products. A shift of dominating electronic configuration will take place, when the system passes through the saddle point (T) region. At one point, the two energies become degenerate and the total wave function is here approximately given as

$$\Psi = \sqrt{\frac{1}{2}}(\Psi_I - \Psi_{II}). \tag{9.15}$$

It is clear that we need a multiconfigurational wave function in order to describe this process properly—single configuration method does not even have the qualitative correct description of this point. Reactions of this type are often called *forbidden* because they result in rather large barriers due to the change of electronic configuration. Other types of reactions are *allowed* because the electron configuration is not changed and as the result the barrier is low. Many reactions of this type occur in biochemical systems where energies cannot be allowed to vary over a large scale.

9.1.5   Other Cases of Near-Degeneracy Effects

The examples of multiconfigurational mixing discussed above refer to situations where one or more chemical bonds are broken, either completely in a dissociation process, or partly at a transition state on an energy surface for a chemical reaction. Also in the ozone case the mixing was due to partial breaking of the OO π-bond in favor of a biradical state. The complete dissociation process leads to the formation of open-shell configurations, while in concerted chemical reactions one closed shell is broken and transformed into another closed shell. In both these cases, multiconfigurational mixing occurs as a result of decoupling paired electrons.

Near-degeneracy effects can occur also in other situations, where different electronic configurations have similar energies.

It is, for example, not unusual to find strong configurational mixing in excited states of molecular systems. Different configurations, singly excited with respect to an RHF ground state, are often close in energy. When they are of the same symmetry, even a small interaction term leads to strong mixing. Typical examples are the alternant hydrocarbons. (These are planar unsaturated molecules, whose carbon atoms can be divided into two sets such that members of one set are formally only π bonded to members of the other set, e.g., benzene, naphthalene, butadiene.) A simple π-electron treatment with, for example, the Pariser–Parr–Pople method yields for these molecules singly excited configurations that occur in pairs with equal energy. A two by two configuration interaction treatment then gives two states (the plus (+) and minus (−) states), which contain the two degenerate configurations with equal weight. All of the transition intensity goes into the plus state, while the transition moment for the minus state is zero in the simplified model. These properties of some of the excited states in alternant hydrocarbons remain approximately valid in more accurate models.

Another important case of near degeneracy occurs in compounds containing transition metal atoms. There are two reasons for strong configurational mixing in such molecules. First, the chemical bonds are often weak, leading to a substantial occupation of the antibonding orbital. The second reason is related to near degeneracies in the transition metal atom. The electronic configurations $d^n s^2$, $d^{n+1} s$, and d^{n+2} often contain spin multiplets with very similar energies. An example is the nickel atom, where the states $d^8 s^2$, ^3F and $d^9 s$, ^3D only differ in energy by 0.6 kcal/mol. The $d^9 s$, ^1D state is located 9.7 kcal/mol above the ^3F ground state, while the energy difference to d^{10}, ^1S is 42.1 kcal/mol. These closely spaced atomic energy levels will influence the chemical bond in nickel compounds considerably, and may lead to strong mixing of different atomic configurations. It is, however, not always the case, since the splitting will be strongly influenced by the ligands, usually stabilizing the d^{10} state at the expense of electronic states where the diffuse $4s$ electron is occupied. It should be emphasized that a correct calculation of the atomic splittings is very difficult and has to include extensive dynamic correlation and relativistic effects (see previous chapters). These features carry over to some extent to the molecular case, and studies of chemical bonds involving one or more transition metal atoms are more complicated than studies of systems containing only main group atoms. These near degeneracies become even more prominent for heavier atoms such as lanthanides and actinides in low oxidation states. The electronic structures of such systems is further complicated by the strong relativistic effects that occur. Electronic states of different spin multiplicity are often mixed due to a strong spin–orbit coupling.

9.1.6 Static and Dynamic Correlation

We have in this section discussed different situations, where the simple single- configurational Hartree–Fock description of the molecular system breaks down. It has been shown that a model can be devised that gives a qualitatively correct wave function in cases where several electronic configurations are close in energy. In the following section, we develop the tools needed to compute such wave functions.

The discussion above has focused on the configurational mixing occurring in near-degenerate systems. One might argue that to account for these effects it is enough to perform a configuration interaction calculation, using a predetermined set of molecular orbitals, obtained for example from an RHF calculation. Thus, there should be no need for a MCSCF theory, which includes orbital optimization. This is, however, a dangerous approach. It is clear that the molecular orbitals will also be strongly affected. If, for example, the MOs for the hydrogen molecule were determined from an SCF calculation, one would not obtain atomic hydrogen orbitals at large internuclear separation. The RHF determinant contains, as was shown in the beginning of this section, a large ionic component. Thus, the orbitals will be intermediate between the orbitals for a hydrogen atom and those of a hydrogen negative ion. Similar errors would occur in more complex systems. For example, multiconfigurational effects often modify the polarity in a chemical bond, leading to considerable modifications of the MOs involved. In many cases, strong correlation effects make the electron density more compact. In general, it is necessary to optimize the orbitals using a qualitatively correct wave function. Thus the need for an MCSCF model.

Correlation effects in molecules are normally partitioned into near-degeneracy effects (static correlation) and dynamic correlation. Qualitatively they differ in the way they separate the electrons. Static correlation leads to a large separation in space of the two electrons in a pair, for example, on two different atoms in a dissociation process. Dynamic correlation, on the other hand, deals with the interaction between two electrons at short interelectronic distance, the cusp region. It should be emphasized that MCSCF methods deal primarily with the near-degeneracy effects. Other methods are used to treat dynamical correlation. These include large-scale configuration interaction methods, coupled cluster methods, and perturbation theory. Such methods will be treated in other parts of the book. There is normally no need for orbital optimization in calculations of dynamical correlation effects, since the electron density is only weakly affected. This is, however, a rule with several exceptions. One concerns the electron structure of transition metal compounds, where dynamic correlation in the $3d$-shell can lead to a substantial modification of the electronic structure, for example, by changing the polarity in chemical bonds. Another example relates to the calculation of the electron affinity of molecules. It may here happen that the extra electron becomes bound only when a large fraction of the dynamic correlation energy is included in the calculation. It follows that the MOs have to be determined on that level of approximation. A similar situation observed for some excited states in molecules, especially those that contain substantial contributions from ionic VB structures in the wave function. A typical example is the V-state of the ethene molecule (the lowest singlet $\pi - \pi^*$ state), which in the single-determinant approximation is completely ionic. In these cases, much too diffuse orbitals are obtained if dynamical correlation effects are not included in the wave function.

The partitioning of the correlation energy into static (near degenerate) and dynamic is in most cases not obvious. Consider, for example, again the H_2 molecule. At large internuclear separations, it is clear that the two configurations Ψ_1 and Ψ_2 in Eq. 9.10 are nearly degenerate. But at shorter distances, there are other configurations that have similar weights as Ψ_2 in the wave function (e.g., $(1\pi_u)^2$), and the effect of Ψ_2 is here better treated as dynamical (left–right) correlation. In practice, this means that we have to define a configuration as near degenerate if it has a large weight somewhere on the section of the energy surface under study. If the MCSCF calculation is used only as a starting point for more elaborate calculations including dynamical correlation effects, this uncertainty is normally not a problem. However, when only MCSCF calculations are performed, serious balance problems may occur, where different portions of the dynamical correlation energy is included on different parts of an energy surface.

Normally, full valence MCSCF calculations (choosing all valence orbitals as active) represent a balanced treatment of correlation. However, this is not always the case, especially not in systems containing lone-pair electrons. For example, a full valence MCSCF calculation for the water molecule yields less accurate values for the bond distance, the bond angle, and the dipole and quadrupole moment than an SCF calculation. The reason is that there are only two orbitals available for correlating the eight valence electrons (the $4a_1$ and the $2b_2$ orbitals, see Figure 9.4). Thus, correlation is only introduced into the lone-pair orbital $3a_1$ and the OH bonding orbital $1b_2$. The orbitals $2a_1$ and $1b_1$ are left uncorrelated, leading to an unbalanced treatment of the correlation of the OH bonds. In fact, if symmetry was relaxed, the molecule would in this approximation become unsymmetrical with two unequal OH bond lengths. The remedy to this problem is to introduce two more active orbitals: $5a_1$ and $2b_1$. It turns out that the occupation numbers for these two orbitals (which do not belong to the valence shell!) is not smaller than those of the correlating orbitals within the valence shell. In addition, much improved results for the properties are obtained. We shall return to the problem of choosing the active orbital space for MCSCF calculations at the end of this chapter.

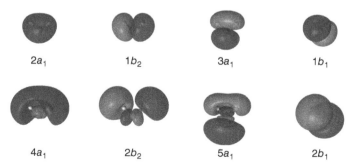

$2a_1$ $1b_2$ $3a_1$ $1b_1$

$4a_1$ $2b_2$ $5a_1$ $2b_1$

Figure 9.4 The valence orbitals of the water molecule plus the $5a_1$ and $2b_1$ extra valence orbitals. The top row orbitals are the bonding orbitals and the second row contains the corresponding correlating orbitals.

Finally, one notices that the definition of the correlation energy becomes less clear-cut when near-degeneracy effects appear in the wave function. Conventionally, the correlation energy has been defined as the difference between the (restricted) Hartree–Fock energy and the exact eigenvalue of the nonrelativistic Hamiltonian. When the RHF energy loses its meaning, so does this definition. The correlation energy for H_2 at large internuclear distance is not 160 kcal/mol; it is zero. It is not enough to use the energy obtained as the sum of the RHF energies for the two dissociation products instead of the energy of the super-molecule, since we need a definition that is valid for the whole potential curve. To use instead the unrestricted Hartree–Fock method as a basis for the definition of correlation energy is not satisfactory. This method does give the right dissociation products but fails to give a qualitatively correct description of the potential curve in the region intermediate between equilibrium and dissociation. Furthermore, the UHF wave function is not an eigenfunction of the total spin. One possibility would be to base the definition on a full valence MCSCF calculation, instead of Hartree–Fock. A full valence MCSCF calculation will always lead to the correct dissociation products. The difference between the corresponding energy and the exact energy could then be used as a definition of the dynamical correlation energy. There are two difficulties with this approach: first it is possible to perform full valence CI calculations only for small systems, containing less than four atoms. Second, since there is no clear-cut partitioning of dynamical and static correlation effects, the definition partly loses its meaning. Obviously, we are here trying to fight a battle with windmills. It is better not to. It is important to understand the effects of electron correlation in different bonding situations and to be able to calculate these effects, when needed. It is possible to manage this without a formal definition of the correlation energy.

9.2   DETERMINATION OF THE MCSCF WAVE FUNCTION

In the examples of the proceeding section, the MCSCF wave functions were compact, comprising only a few CSFs. This is, however, not the standard case as we shall see later. The MC expansion can be long, containing thousand or in extreme cases even millions of CSF. In this section, we see how one can determine the CI expansion coefficients and the molecular orbitals for such wave functions.

Let us start by writing down the MCSCF wave function:

$$|0\rangle = \sum_m c_m |m\rangle, \tag{9.16}$$

where $|0\rangle$ is the MCSCF wave function, c_m the expansion coefficients, and $|m\rangle$ the CSFs or Slater determinants. We can use either as is discussed below. The CSFs have the virtue of giving a compact expansion where each term is also a spin eigenfunction. Slater determinants, on the other hand, give simpler matrix elements of the

Hamiltonian operator and is therefore in modern programs often used to solve the often large eigenvalue problem. The Hamiltonian is given as

$$\hat{H} = \sum_{i,j} h_{ij}\hat{E}_{ij} + \frac{1}{2}\sum_{i,j,k,l} g_{ijkl}(\hat{E}_{ij}\hat{E}_{kl} - \delta_{jk}\hat{E}_{il}). \tag{9.17}$$

For a normalized CI wave function of the type 9.16, we obtain the energy as the expectation value of the Hamiltonian 9.17:

$$E = \langle \Psi | \hat{H} | \Psi \rangle = \sum_{i,j} h_{ij}D_{ij} + \sum_{i,j,k,l} g_{ijkl}P_{ijkl}. \tag{9.18}$$

This energy expression forms the basis for the derivation of the MCSCF optimization methods. Note that the information about the molecular orbitals (the MO coefficients) is contained completely within the one- and two-electron integrals. The density matrices D and P contain the information about the CI coefficients.

9.2.1 Exponential Operators and Orbital Transformations

In this section, we go through some of the formalism needed for the coming derivation of the optimization methods. The parameters to be varied in the energy expression 9.18 are the CI coefficients and the molecular orbitals. We consider these variations as rotations in an orthonormalized vector space. For example, variations of the MOs correspond to a unitary transformation of the original MOs into a new set:

$$\boldsymbol{\varphi}' = \boldsymbol{\varphi}\boldsymbol{U}, \tag{9.19}$$

where $\boldsymbol{\varphi}$ is a row vector containing the original orbitals, and $\boldsymbol{\varphi}'$ is the transformed orbital vector. $\boldsymbol{U}$ is a unitary matrix:

$$\boldsymbol{U}^{\dagger}\boldsymbol{U} = \boldsymbol{1}. \tag{9.20}$$

The corresponding transformation of the spin-orbitals is obtained by multiplying Eq. 9.19 with an α or β spin function. When we make the transformation from one set of spin-orbitals to the other, the annihilation and creation operators will change. The following relations are easily established by operating with the creation operator in the primed space on the vacuum state:

$$\hat{a}'_i = \sum_j \hat{a}_j U^*_{ji},$$

$$\hat{a}'^{\dagger}_i = \sum_j \hat{a}^{\dagger}_j U_{ji}. \tag{9.21}$$

Alternatively, it is possible to write the transformed annihilation and creation operators in the following form:

$$\hat{a}'_i = e^{-\hat{T}}\hat{a}_i e^{\hat{T}},$$

$$\hat{a}'^{\dagger}_i = e^{-\hat{T}}\hat{a}^{\dagger}_i e^{\hat{T}}, \tag{9.22}$$

where $\hat{T}$ is an anti-Hermitian operator:

$$\hat{T} = \sum_{i,j} T_{ij} \hat{a}_i^\dagger \hat{a}_j \tag{9.23}$$

with the matrix T anti-Hermitian: $T^\dagger = -T$. The proof of Eqs. 9.22 is obtained by Taylor expanding the operators on both sides and collect terms with the same order into anticommutators:

$$\hat{a}_i'^\dagger = \hat{a}_i^\dagger + [\hat{a}_i^\dagger, \hat{T}] + \frac{1}{2}[[\hat{a}_i^\dagger, \hat{T}], \hat{T}] + \cdots \tag{9.24}$$

and a similar relation for the annihilation operator. The commutators in Eq. 9.24 are evaluated using the anticommutator rules, for example:

$$[\hat{a}_i^\dagger, \hat{T}] = \sum_{k,l} (\hat{a}_i^\dagger \hat{a}_k^\dagger \hat{a}_l - \hat{a}_k^\dagger \hat{a}_l \hat{a}_i^\dagger) T_{kl} = -\sum_k \hat{a}_k^\dagger (T)_{ki},$$

$$[[\hat{a}_i^\dagger, \hat{T}], \hat{T}] = \cdots = \sum_k \hat{a}_k^\dagger (T^2)_{ki}. \tag{9.25}$$

We finally obtain by summing over all terms in Eq. 9.24:

$$e^{-\hat{T}} \hat{a}_i^\dagger e^{\hat{T}} = \sum_k \hat{a}_k^\dagger (1 - T + \frac{1}{2} T^2 + \cdots) = \sum_k \hat{a}_k^\dagger (e^{-T})_{ki}. \tag{9.26}$$

We can now identify the unitary matrix: $U = e^{-T}$ (note that e^{-T} is a matrix and $e^{-\hat{T}}$ is an operator). It is a general property of unitary matrices that they can be written as the exponential of an anti-Hermitian matrix: First, it is immediately clear that e^{-T} is a unitary matrix, when T is anti-Hermitian. Second, it is possible to show that all unitary matrices can be written in this form.

We have thus shown the relations (see Eq. 9.22) for the transformation of the annihilation and creation operators to a new spin-orbital basis. We can use these relations to express an arbitrary Slater determinant in the new basis in terms of the determinants in the original basis. In order to do so, we generate the Slater determinant by applying a sequence of creation operators on the vacuum state:

$$|m'\rangle = \hat{a}_i'^\dagger \hat{a}_j'^\dagger \hat{a}_k'^\dagger \cdots |vac\rangle = e^{-\hat{T}} \hat{a}_i^\dagger e^{\hat{T}} e^{-\hat{T}} \hat{a}_j^\dagger e^{\hat{T}} \cdots |vac\rangle$$

$$= e^{-\hat{T}} \hat{a}_i^\dagger \hat{a}_j^\dagger \hat{a}_k^\dagger \cdots |vac\rangle = e^{-\hat{T}} |m\rangle. \tag{9.27}$$

This is a very important relation. It shows that the effect of an orbital transformation on a Slater determinant can be obtained simply by operating on that determinant with the exponential operator $e^{-\hat{T}}$. We shall make extensive use of this equation in the derivation of the MCSCF optimization schemes.

So far we have considered an arbitrary unitary transformation of the spin-orbitals. In practice, we shall only transform the spatial part of the orbitals: the molecular orbitals. In order to see the implication of this for the operator $\hat{T}$, we order the spin-orbitals by multiplying first the MOs, φ, with an α spin function, followed by the spin-orbitals obtained by multiplying the same MOs with a β spin function:

$$\phi = (\varphi\alpha, \varphi\beta). \tag{9.28}$$

The matrix T can then be arranged as a collection of four submatrices corresponding to transformations within and between the two spin-orbital blocks:

$$T = \begin{pmatrix} T_{\alpha\alpha} & T_{\alpha\beta} \\ T_{\beta\alpha} & T_{\beta\beta} \end{pmatrix}. \tag{9.29}$$

Now, $T_{\alpha\beta} = T_{\beta\alpha} = 0$, since we are not mixing orbitals with different spin. Furthermore, the transformation matrix is the same for α and β. Thus, we also have $T_{\alpha\alpha} = T_{\beta\beta}$. The $\hat{T}$ operator can now be summed over the molecular orbitals as

$$\hat{T} = \sum_{i,j} (T_{ij}^{\alpha\alpha} \hat{a}_{i\alpha}^\dagger \hat{a}_{j\alpha} + T_{ij}^{\alpha\beta} \hat{a}_{i\alpha}^\dagger \hat{a}_{j\beta} + T_{ij}^{\beta\alpha} \hat{a}_{i\beta}^\dagger \hat{a}_{j\alpha} + T_{ij}^{\beta\beta} \hat{a}_{i\beta}^\dagger \hat{a}_{j\beta}). \tag{9.30}$$

If we now use the relation given above for the matrix T, we obtain

$$\hat{T} = \sum_{i,j} T_{ij}(\hat{a}_{i\alpha}^\dagger \hat{a}_{j\alpha} + \hat{a}_{i\beta}^\dagger \hat{a}_{j\beta}) = \sum_{i,j} T_{ij}\hat{E}_{ij}, \tag{9.31}$$

where we have introduced the excitation operators $\hat{E}_{ij}$ and dropped the labels $\alpha\alpha$ $(\beta\beta)$ for the matrix elements of T. T_{ij} is an element of this anti-Hermitian matrix and describes the unitary rotation of the MOs through $U = e^{-T}$. We shall consider only real molecular orbitals. Thus, the matrix T is real and antisymmetric ($T_{ij} = -T_{ji}$, $T_{ii} = 0$). We can use this to rewrite Eq. 9.31 (note the restricted summation) as

$$\hat{T} = \sum_{i>j} T_{ij}(\hat{E}_{ij} - \hat{E}_{ji}) = \sum_{i>j} T_{ij}\hat{E}_{ij}^-. \tag{9.32}$$

An orthogonal rotation is thus described by the "replacement" operators $\hat{E}_{ij} - \hat{E}_{ji}$.

9.2.2 Slater Determinants and Spin-Adapted State Functions

So far we have based the formalism on an N-electron basis built from Slater determinants. However, as shown above, both the Hamiltonian and the orbital rotations can be described in terms of the orbital excitation operators $\hat{E}_{ij}$. All matrix elements involving the Hamiltonian and the operator $\hat{T}$ can therefore be computed without explicit reference to the spin-orbital basis. We note, in addition, that the excitation operators commute with the spin operators $\hat{S}_z$ and $\hat{S}^2$. It is then possible to work entirely in a spin-adapted configurational basis. The total MCSCF wave function will be assumed to be a pure spin eigenfunction. It therefore seems favorable to be able to expand it in terms of a spin-adapted configurational basis. There are many ways in which spin-adapted configurations can be generated from a basis of Slater determinants. One of the most commonly used methods today is the graphical unitary group approach (GUGA).

Most of the formalism to be developed in the coming sections of these notes will be independent of the specific definition of the configurational basis, in which we expand the wave function. We therefore do not have to be very explicit about the exact nature of the basis states $|m\rangle$. They can be either Slater determinants or the spin-adapted CSF. For a long time, it was assumed that CSFs were to be preferred for MCSCF calculations, since it gives a much shorter CI expansion. Efficient methods such as GUGA had also been developed for the solution of the CI problem. Recent experience has, however, shown that calculations in a Slater determinant basis can be performed very efficiently on modern computers. In fact, the largest CI calculations carried through to date have been performed using Slater determinants (the limit for full CI calculations is today of the order of 10^{10} Slater determinants).

The variational parameters for the CI part of the wave function could be taken to be the CI coefficients C_m in the expansion of the MCSCF wave function, which we now write as

$$|0\rangle = \sum_m C_m |m\rangle, \tag{9.33}$$

where we do not here specify the exact nature of the basis states $|m >$, other than requiring orthonormality. The variation of the CI coefficients are made with the restriction that the total wave function, Eq. 9.33, remains normalized:

$$\sum_m |C_m|^2 = 1. \tag{9.34}$$

We can remove the problem with this subsidiary condition by instead using as the variational space the orthogonal complement to the MCSCF state $|0\rangle$. This variational space is defined as a set of states $|K\rangle$ expanded in the same set of basis states $|m\rangle$ as $|0\rangle$:

$$|K\rangle = \sum_m C_m^K |m\rangle, \tag{9.35}$$

with the property $\langle K|L\rangle = \delta_{KL}$. To each of the states $|K\rangle$ corresponds a variational parameter, describing the contribution of this state to a variation of the MCSCF state $|0\rangle$. This variation can be described as a unitary rotation between the MCSCF state and the complementary space. The operator that performs this rotation is constructed in the same way as was done for the orbital rotations. We start by defining an anti-symmetric replacement operator:

$$\hat{S} = \sum_{K \neq 0} S_{K0}(|K\rangle\langle 0| - |0\rangle\langle K|). \tag{9.36}$$

The corresponding unitary operator is $e^{\hat{S}}$. In the following section, we use these operators in a derivation of the MCSCF optimization methods.

9.2.3   The MCSCF Gradient and Hessian

In this section, we compute the energy gradient and the Hessian matrix corresponding to the energy expression, Eq. 9.18 . We introduce the variation of the CI coefficients by operating on the MCSCF state $|0\rangle$ with the unitary operator $e^{\hat{S}}$, with $\hat{S}$ defined by Eq. 9.36 and the orbital rotations by the operator $e^{\hat{T}}$, with $\hat{T}$ from Eq. 9.32 . A variation of the MCSCF state can thus be written as

$$|0'\rangle = e^{\hat{T}} e^{\hat{S}} |0\rangle. \tag{9.37}$$

The order of the operators in Eq. 9.37 is *not* arbitrary because they do not commute. The reverse order, however, leads to more complicated expressions for the Hessian matrix, and as the final result is invariant of the order, we make the more simple choice given in Eq. 9.37 . The energy corresponding to the varied state will be a function of the parameters in the unitary operators, and we can calculate the first and second derivatives of this function by expanding the exponential operators to second order. The unitarity of the operators guarantees that $|0'\rangle$ remains normalized if the original state $|0\rangle$ is normalized. The energy is obtained as a function of the rotation parameters T and S as

$$E(T, S) = \langle 0 | e^{-\hat{S}} e^{-\hat{T}} \hat{H} e^{\hat{T}} e^{\hat{S}} | 0 \rangle. \tag{9.38}$$

If we expand the exponential operators to second order, we obtain for the operator in Eq. 9.38 the following expression:

$$\begin{aligned} E(T, S) = \langle 0 | \hat{H} \\ + [\hat{H}, \hat{T}] + [\hat{H}, \hat{S}] \\ + \frac{1}{2} [[\hat{H}, \hat{T}], \hat{T}] + [[\hat{H}, \hat{S}], \hat{S}] + [[\hat{H}, \hat{T}], \hat{S}] + \cdots | 0 \rangle. \end{aligned} \tag{9.39}$$

The first term in this expression is the zeroth-order energy $E(0, 0)$. The two next terms give the first derivatives with respect to the parameters T_{ij} (Eq. 9.32) and S_{K0} (Eq. 9.36). We obtain the following result for the first derivative with respect to the orbital rotation parameters:

$$\langle 0 | [\hat{H}, \hat{T}] | 0 \rangle = \sum_{i>j} T_{ij} \langle 0 | [\hat{H}, \hat{E}_{ij}^-] | 0 \rangle, \tag{9.40}$$

which gives the derivative as

$$g_{ij}^o = \langle 0 | [\hat{H}, \hat{E}_{ij}^-] | 0 \rangle. \tag{9.41}$$

Here the superscript o has been used to indicate that this is the derivative with respect to the orbital rotation parameters. Note the close resemblance between Eq. 9.41 and the corresponding expression in Hartree–Fock theory—the Brillouin theorem. It is therefore sometimes called the *Extended (or generalized) Brillouin Theorem*. The Brillouin theorem states that there is no interaction between the Hartree–Fock wave function and singly excited configurations. Equation 9.41 yields the corresponding condition for an MCSCF wave function: the matrix element on the right-hand side of the equation is zero for optimized orbitals. If $|0\rangle$ is an Hartree–Fock state, the only interesting matrix elements are those where j corresponds to an occupied and i to a virtual orbital. Setting the derivative equal to zero for these matrix elements immediately leads to the Hartree–Fock equations. Note that, if i and j correspond to empty orbitals, or if both orbitals are doubly occupied, 9.41 is identically zero—this trivially due to the nature of the operator $\hat{E}_{ij}^-$.

The situation in the MCSCF case is obviously more complex because orbitals may be partly occupied—they are occupied in some terms in the wave function but not in others. However, if the occupation number of both orbitals are exactly equal to two, Eq. 9.41 equals again zero. Like the SCF energy, the MCSCF energy is also invariant toward rotations among the inactive (doubly occupied) orbitals. Rotations between them should thus not be included. The same is obviously true in the case where both occupation numbers are zero (virtual orbitals).

To arrive at the derivatives with respect to the CI parameters, we compute the matrix element of the third term in Eq. 9.39:

$$\langle 0|[\hat{H}, \hat{S}]|0\rangle = \sum_{K\neq 0} S_{K0}(\langle 0|\hat{H}|K\rangle + \langle K|\hat{H}|0\rangle) \tag{9.42}$$

we obtain for real wave functions the derivative (the superscript c has been used to indicate that it is the derivative with respect to the CI parameters):

$$g_K^c = 2\langle 0|\hat{H}|K\rangle. \tag{9.43}$$

This equation tells us that an optimized MCSCF state (for which the derivative is zero) will not interact with the orthogonal complement, that is, it is a solution to the secular problem:

$$(\boldsymbol{H} - E\boldsymbol{1})\boldsymbol{C} = \boldsymbol{0}, \tag{9.44}$$

where $\boldsymbol{H}$ is the Hamiltonian matrix, $\boldsymbol{C}$ the expansion coefficients of the MCSCF wave function, and E the energy. This is the condition we expected to obtain for an optimized CI wave function.

This leads us to the third orbital invariance for the fully optimized CASSCF wave function. In the case that both the orbital indices of the $\hat{E}_{ij}^-$ operator refer to active orbitals, we have that $\hat{E}_{ij}^-|0\rangle = \sum_{K\neq 0} C_K|K\rangle$; this translates to that Eq. 9.41 is expressed as

$$g_{ij}^o = \sum_{K\neq 0} C_K(\langle 0|\hat{H}|K\rangle + \langle K|\hat{H}|0\rangle) \tag{9.45}$$

However, since for an optimized CASSCF wave function we have that Eq. 9.43 equals zero, we also see that g_{ij}^o is zero. This concludes the orbital invariance of the optimized CASSCF wave function to be established for inactive–inactive, active–active, and virtual–virtual rotations. Note that while the inactive–inactive and virtual–virtual invariances are always fulfilled, the active–active invariance is only fulfilled for the optimized CASSCF wave functions. However, as long as the

CI-expansion coefficients are optimized in each step of the optimization procedure, the active–active rotation is redundant and should be ignored.

In similar way to the gradients, we can obtain the second derivatives of the wave function (the Hessian matrix) with respect to the variational parameters. We shall not pursue this derivation in detail here. The interested reader can find the details in, for example, Ref. [1]. A few words, however, about the optimization strategy. In order to do so, we introduce a common parameter p and write the energy expression, Eq. 9.39, as

$$E(p) = E(p = 0) + \sum_i \left(\frac{\partial E}{\partial p_i} \right)_{p=0} p_i + \sum_{i,j} p_i \left(\frac{\partial^2 E}{\partial p_i \partial p_j} \right)_{p=0} p_j + \cdots \quad (9.46)$$

or in matrix notation:

$$E(p) = E(0) + g^\dagger p + p^\dagger H p + \cdots . \quad (9.47)$$

Here we have defined the energy gradient vector g and the Hessian matrix H with the elements given as

$$g_i = \left(\frac{\partial E}{\partial p_i} \right)_{p=0} \quad \text{and} \quad H_{ij} = \left(\frac{\partial^2 E}{\partial p_i \partial p_j} \right)_{p=0} . \quad (9.48)$$

The stationary points on the energy surface, Eq. 9.46, are obtained as solutions to the equations $\partial E / \partial p_i = 0$. They can be approximately solved by starting from the expansion, Eq. 9.47, truncated at second order. Setting the derivative of E in Eq. 9.47 equal to zero leads to the system of linear equations:

$$g + Hp = 0 \quad \text{or} \quad p = -H^{-1}g. \quad (9.49)$$

This is the Newton–Raphson (NR) procedure. A new point is obtained by solving Eq. 9.49, redefining the solution as the new zero point, recalculating g, and H and returning to Eq. 9.49 . Such a procedure converges quadratically, that is, the error vector in iteration n is a quadratic function of the error vector in iteration $n - 1$. This does not necessarily mean that the NR procedure will converge fast, or even at all. However, close to the stationary point, we can expect a quadratic behavior. There are many ways in which these equations can be solved and we refer to Ref. [1] for more details. The difficulty to compute the Hessian matrix, in particular the coupling terms between orbital and CI rotations, has led to a variety of simplified approaches, where the Hessian matrix is approximated. We leave these problems here and shall instead move to the more chemically interesting problem of how to construct a viable MCSCF wave function.

9.3 COMPLETE AND RESTRICTED ACTIVE SPACES, THE CASSCF AND RASSCF METHODS

Programs are available today that perform SCF (closed shell or unrestricted) calculations on molecules almost automatically. The user only has to provide the structural information about the molecule, and maybe suggest a basis set. This is an important development, since the user can then concentrate on his chemical problem, and does not have to know much about the program he is running. However, it is important to know what one is doing and what can be expected to be the outcome of a calculation. This means that some feeling has to be acquired for the relation between basis set and accuracy, and the size of the correlation corrections for the type of problem under study. This knowledge cannot, of course, be exact, but is rather a collection of experiences from earlier work and from calibration studies. A prerequisite for responsible studies of molecules using the SCF method is therefore a reasonably thorough knowledge of earlier work in the field.

Many quantum chemists today use density functional theory (DFT). The comments made above are equally valid here with the additional problem that one should be aware of the accuracy that may be obtained using different functionals. This accuracy may vary widely from one type of molecules to another and for different molecular properties. Again, it is important to use the collected experience in judging the quality of the results.

The "black box" situation of SCF and DFT applications has not yet been reached for MCSCF theory. This constitutes a major problem, since MCSCF is a much better starting point for quantum chemical calculations on many interesting chemical problems (a good example is studies of transition states for chemical reactions). A development toward more automatized procedures for this approach would therefore be of great value. In spite of several attempts, this has however turned out to be a difficult task.

Why is it then more complicated to perform an MCSCF calculation? The basis set problem is similar in SCF and MCSCF theory, and can be solved by the construction of basis set libraries containing well-documented and thoroughly tested basis sets of varying quality. The major difference lies in the construction of the wave function. In SCF or DFT theory, it is given as a single determinant, where the occupied orbitals are determined by the optimization procedure. This is not so in MCSCF theory. Here a decision on the general structure of the wave function has to be made beforehand. This cannot be done without an a priori knowledge about the electronic structure. In normal bonding situations, such a knowledge is not too hard to achieve. A few examples were given in an earlier Section 9.1.1. However, in more complex situations—such as the transition state for a chemical reaction—it may be difficult to make a priori judgments about the most important electronic configurations to include in the MCSCF wave function. Below we discuss how this problem can be, at least partly, solved by the Complete Active Space Method. Here, the problem is reduced to defining a set of active orbitals, which describe the near-degeneracy effects. The choice of active orbitals requires an insight into the electronic structure, which is often rather obvious, but, not always. There are many cases where the choice is not at all clear, and several trials have to be made before the best choice has been found. This is far from black box situation, and the procedure is not easily automatized.

The CASSCF method is based on a partitioning of the occupied molecular orbitals into subsets, corresponding to how they are used to build the wave function. We define for each symmetry block of MOs the following subsets:

1. Inactive orbitals
2. Active orbitals
3. External orbitals.

The inactive and active orbitals are occupied in the wave function, while the external (also called secondary or virtual) orbitals span the rest of the orbital space, defined from the basis set used to build the molecular orbitals. The inactive orbitals are kept doubly occupied in all configurations that are used to build the CASSCF wave function. The number of electrons occupying these orbitals is thus twice the number of inactive orbitals. The remaining electrons (called active electrons) occupy the active orbitals. We illustrate the active spaces for one CAS and one RAS wave function in Figure 9.5. The left part of the figure shows a component of a CAS wave function with two electrons in three orbitals coupled to a singlet.

The CASSCF method is an attempt to generalize the Hartree–Fock model to situations where near degeneracies occur, while keeping as much of the conceptual simplicity of the RHF approach as possible. Technically, the CASSCF model is by necessity more complex, since it is based on a multiconfigurational wave function. The building blocks are, as in the RHF model, the occupied (inactive and active) orbitals. The number of electrons is, however, in general less than twice the number of occupied orbitals. The number of electron configurations generated by the orbital

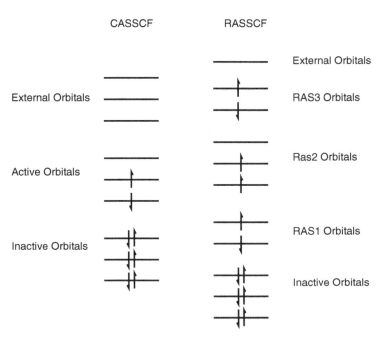

Figure 9.5 Illustration of the active orbitals for a 3-in-2 CAS, extended to a RAS of SD type.

space is therefore larger than unity. The total wave function is formed as a linear combination of all the configurations, in the N-electron space, that fulfill the given space and spin symmetry requirements and have the inactive orbitals doubly occupied. It is in this sense "complete" in the configurational space spanned by the active orbitals. The inactive orbitals represent an "SCF sea" in which the active electrons move. These orbitals have occupation numbers exactly equal to two, while the occupation numbers of the active orbitals varies between zero and two. It is obvious that the inactive orbitals should be chosen as the orbitals that are not expected to contribute to near-degeneracy correlation effects.

The conceptual simplicity of the CASSCF model lies in the fact that once the inactive and active orbitals are chosen, the wave function is completely specified. In addition, such a model leads to certain simplifications in the computational procedures used to obtain optimized orbitals and CI coefficients, as was illustrated in the preceding sections. The major technical difficulty inherent to the CASSCF method is the size of the complete CI expansion, N_{CAS}. It is given by the so-called Weyl formula, which gives the dimension of the irreducible representation of the unitary group $U(n)$ associated with n active orbitals, N active electrons, and a total spin S (note the use of so-called binomial coefficients):

$$N_{CAS} = \frac{2S + 1}{n + 1} \binom{n + 1}{N/2 - S} \binom{n + 1}{N/2 + S + 1}. \tag{9.50}$$

Obviously, N_{CAS} increases strongly as a function of the size n of the active orbital space. In practice, this means that there is a rather strict limit on the size of this space. Experience shows that this limit is normally reached for n around 12–16 orbitals, except for cases with only a few active electrons or holes. The large number of CASSCF applications performed to date illustrates clearly that this limitation does not normally create any serious problem. It should be remembered that the CASSCF model is an extension of the RHF scheme. As such, it is supposed to produce a good zeroth-order approximation to the wave function when near degeneracies are present. This goal can in most cases be achieved with only a few active orbitals. The CASSCF method has not been developed for treating dynamical correlation effects but to provide a good starting point for such studies.

However, there exist cases where it is advantageous to be able to use a larger set of active orbitals. The dimension of the CAS wave function can then become prohibitively large, and it may be of interest to look for other means of restricting the expansion length. One possibility is to extend the CAS partitioning of the orbital space to a more restricted form (the RASSCF method) that can hold many more active orbitals. This is done by introducing to new orbital partitionings, RAS1 and RAS3. RAS1 is part of the inactive space, but here one or more holes are allowed. Electrons may be excited to the other subspaces. RAS3 is part of the virtual space and is allowed to hold one or more electrons, that is, electrons may be excited into this space. The original active space is now called RAS2. Such a wave function is illustrated by the right-hand part of Figure 9.5 that shows a case with two electrons in RAS2 coupled to a triplet, two holes in RAS1 and two electrons in RAS3. In the following section, we discuss the choice of active orbitals for different type of systems and molecular processes.

9.3.1 State Average MCSCF

One of the most common applications of the CASSCF and RASSCF methods is for studies of excited states for electron spectroscopy, photophysical, and photochemical processes. To perform RASSCF calculations on excited states is, however, often difficult and can lead to convergence problems, root flipping, and other difficulties that are problematic to handle. One approximation, which usually works well, is to perform state average calculations, where the orbitals are optimized for a number of electronic states. The energy is written as an average of the energy for each individual state:

$$E^{av} = N^{-1} \sum_i^N E_i, \qquad (9.51)$$

where N is the number of considered states. The energy can be written in terms of averaged density matrices (compare 9.18):

$$E^{av} = \sum_{i,j} h_{ij} D_{ij}^{av} + \sum_{i,j,k,l} g_{ijkl} P_{ijkl}^{av}, \qquad (9.52)$$

where the average densities are easily obtained from the densities of the individual states. The same code can thus be used as for single state calculations by just replacing the density matrices with the average values.

State average calculations are necessary in cases where several states are close in energy or when the target of the simulation for other reasons involve several states as in optical spectroscopy, photochemistry, and photophysics. These typically included avoided crossings or conical intersections along a reaction coordinate. Another typical case is in electron spectroscopy when valence excited and Rydberg states are close in energy. We shall encounter several examples in the application part.

9.3.2 Novel MCSCF Methods

In light of the exponential scaling of the CASSCF and RASSCF models with respect to the size of the active space, several groups are today trying to overcome these limitations. These new methods have been designed to be able to effectively apply the CASSCF and RASSCF methods to chemical systems that are for practical reasons beyond present day's limitations (today a conventional CASSCF calculation beyond 16 electrons in 16 orbitals defines this bottleneck, although several examples of conventional CASSCF in the literature have treated larger active spaces). These new methods are (i) the density matrix renormalization group (DMRG) approach [2],[3], (ii) the so-called variational calculation of the two-electron reduced density matrix (2-RDM) by Mazziotti [4], (iii) the graphical-contracted-function (GCF) MCSCF method [5], and (iv) the perfect quadruples method [6]. These methods' current status is that they are far from perfect and that more developments and improvements are needed to have the methods to achieve some of the properties associated with the current conventional methods; some such properties are unique solutions, invariance to orbital rotations within a given orbital subspace, robust convergence techniques, and ability to compute energies and properties of excited states. Despite these initial problems, the new approaches have in practical applications been able to demonstrate results that has brought new insights to known chemical problems. We expect to see more of these methods in the near future; ongoing research will make the current practical problem something of the past. Note, however, the success of these new methods will not in any way alter how the CASSCF and RASSCF methods will be used; the only difference is that larger systems as compared to today can be studied.

For a recent review on the state of the art of new parameterizations of the wave functions, as for example the matrix products states (MPS), tree tensor networks (TTN) or projected entangled pair states (PEPS), we recommend the reading of the review by Chan [7].

9.4   CHOOSING THE ACTIVE SPACE

Every application poses its own set of constraints in the choice of the active space, so it is impossible to set up a general procedure for the choice. Such attempts have been made in order to make the CASSCF method more like a "black box" procedure in line with other quantum chemical methods, but they have failed. There are simply too many aspects that must be considered. Below we demonstrate how one can choose the active space in some typical cases. As you shall see, the different parts of the periodic system poses different problems that need to be considered. Before we proceed, just a brief comment on the selection of orbitals for the active space. To do that, it is necessary to ask what is the purpose of the various steps in the CASSCF/CASPT2 approach. The CASSCF step is supposed to bring in an accurate qualitative understanding of the chemical problem at hand, while the subsequent use of the CASPT2 method is expected to give a quantitative accuracy to the modeling. In this aspect the design of the active space is primarily a matter of ensuring a qualitative accuracy. However, we have to live with the fact that the CASPT2 method is just a second-order scheme and sometimes it does not provide us with the desired quantitative accuracy. In this respect, on some occasions some dynamical electron correlation has to be included in the active space. Hence, sometimes the active space is enlarged not to improve the qualitative description of the chemistry but to facilitate the subsequent CASPT2 method to provide us with the desired quantitative accuracy. Below we see examples of both cases.

9.4.1   Atoms and Atomic Ions

For atoms, it is usually rather straightforward to choose the active space. Normally, it is sufficient to have the valence orbitals active, perhaps with added Rydberg-type orbitals for studies of excited states. For second-row atoms, this means the 2s and

Periodic table of the elements

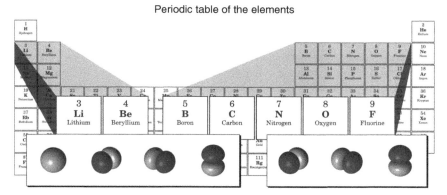

Figure 9.6 The standard active space selection for second-row atoms. For the initial set of elements, Li-C, the 2s and 2p near degeneracy has to be considered. However, for N and onward the 2s orbital can be left inactive.

Periodic table of the elements

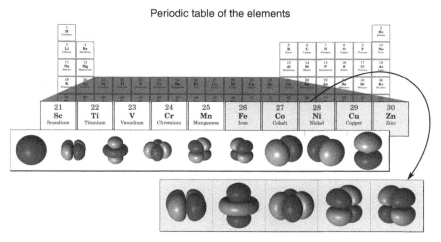

Figure 9.7 The standard active space selection for transition metals. For the first-row transition metals we have 4s, 3d, and 4p as active orbitals. For elements with more than five d-electrons, a second complete 3d'-shell might be required. This effect decreases for higher row transition metals.

2p orbitals, but if there are more than four valence electrons (N–F), one can usually leave the 2s orbital inactive (see Figure 9.6). The same applies for the heavier main group atoms.

Transition metals pose an additional problem (see Figure 9.7). For the first-row transition metals, the obvious active space contains the 4s, 3d, and 4p orbitals, as dictated by the qualitative understanding of the degeneracy effects for these atoms. However, if the 3d-shell is more than half full, one may need an extra 3d'-shell to account for the strong dynamic correlation effects (the so-called double-shell effect) that occur in first- row transition metals. This is a case in which the active space is enlarged due to quantitative accuracy requirements. The double-shell effects are less severe for the second and third row and the extra orbital set is then not needed. It should be mentioned here that requesting chemical accuracy (1–2 kcal/mol accuracy) for a transition metal complex can require additional extensions of the active space, in particular, the inclusion of the 3s and 3p in the active space—semicore correlation. Lanthanide atoms and ions (see Figure 9.8) need the active space 6s, 4f, and 5p, in all 11 orbitals, and 5d, in all 16 orbitals, in the worst case. But the situation is usually simplified in complexes, where the atom is in a high oxidation state.

Actinides represent the most complicated case (see Figure 9.9). Here, the orbitals 7s, 5f, 6d, and 7p have similar energies and are occupied in low lying electronic states. One should therefore ideally use 16 active orbitals. Again, the situation is simplified in higher oxidation states, where 5f dominates. A similar double-shell effect, but now for the f orbitals, could also be expected for these two rows; however, again we would expect the effect to be of lesser importance for the second row (the actinides).

9.4.2   Molecules Built from Main Group Atoms

Ideally, the rules applied to the atoms and atomic ions should be applied to the molecular cases too; however, this would often lead to, if not very small molecular systems are studied, that the active space is larger than what is practically possible to handle today. Hence, some additional rules have to be exercised to limit the active space. Note, still in a future when the next generation of MCSCF algorithms allows much

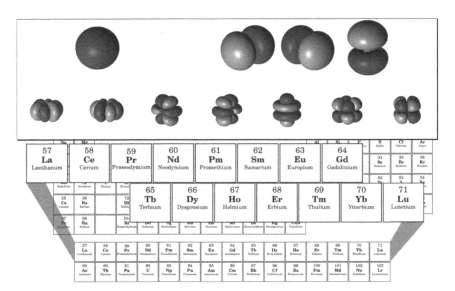

Figure 9.8   The standard active space selection for lanthanides. For the lanthanides, we have 6s, 4f, 5d, and 6p as active orbitals. The 5d-shell normally only is required in the worst case.

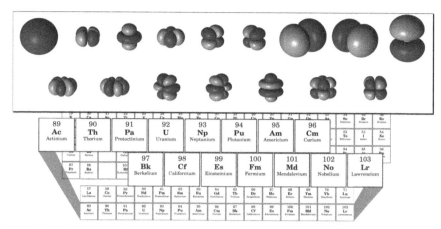

Figure 9.9 The standard active space selection for actinides. For the actinides, we have 7s, 5f, 6d, and 7p as active orbitals.

larger active spaces rules such as these will be useful and of significant practical importance.

Let us first try to identify a general rule to have in mind. The reason why we are using MCSCF theory is that we have situations in which the wave function is multiconfigurational in nature. From the first section of this chapter, we learned that these situations will occur when at least two different electronic structures are close in energy. The literature is by now full of cases that document this and from these we have learnt that two (or several) such configurations normally differ by a one (single) double excitation. The pair of orbitals that constitute the difference between the two electronic structures or configurations are denoted by the correlating pair and in general it is of importance to identify these pairs and include all of them in the active space. In addition to this general rule, which deals directly with why the multiconfigurational approach is needed in the first place, the computation of electronic spectra require that additional orbitals are included in the active space and if open-shell systems are studied that the orbitals of these singly occupied orbitals are included in the active space too. Examples of correlating pairs are, for example, the π and π^* orbitals of conjugated systems and the σ and σ^* orbitals of a bond during bond breaking/formation. Lone-pair orbitals normally do not have a correlating partner and could be left out of the active space if not part of a conjugated system or if one is not specifically interested in the excitation of a lone-pair electron.

Furthermore, as a rule, it is wise to carefully consider what kind of chemical reaction or process is the subject of the study. This can, most of the time, lead to that the same molecular system can be studied with different active spaces. We can identify these different processes as (i) electronic spectroscopy and (ii) dissociative or associative chemical reactions (bond formation and/or breakage). For the first case, it is necessary to identify from and to which orbitals the electrons are promoted (core orbital, valence orbital, Rydberg orbitals, etc.). These active spaces typically do not include any σ type of orbitals since bond breakage/formation is not a part of the study.

In the latter case, it is important to identify all possible correlating pairs of orbitals along the reaction path. Finally, before we start to give some explicit example, we would like to give one more condition to consider. In a bond formation process, it is typical that both the bonding and antibonding orbital of this particular bond is included in the active space. If this formed bond has one or several equivalent partners in the final product, then the bonding and antibonding orbitals of these bonds need to be included in the active space too. If not, the treatment will overemphasize the antibonding character of the newly formed bond while in the already existing equivalent bonds the model with be biased toward bonding. Such an imbalance should be avoided since the results could be erratic.

Here follows a short list of some additional suggestions that can help:

- CH bonds can often be left inactive.
- Orbitals of active end groups on long alkyl chains need to be included while remaining orbitals can be treated as inactive.
- For excited states of planar unsaturated molecules, include all π orbitals if possible. Otherwise, select by energy criteria.
- When studying electronic spectra including Rydberg states (above 5 eV for second-row elements) include the Rydberg orbital(s) in the active space. Note that specially designed Rydberg basis sets should be employed. The addition of just more diffuse functions to the atomic basis set spells disaster.

In Chapter 13, there are a number of chemical cases, covering elements from the whole periodic table, which are studied within the CASSCF method These notes contain specific remarks and motivations on the choice of the active space in every given case. We urge the keen reader to study this chapter in more detail to gain insight into the considerations that must be regarded in order to be able to perform CASSCF calculations so that the predictive power of the method can be utilized.

9.5   REFERENCES

[1] Widmark PO, editor. European Summer School in Quantum Chemistry. vols I-III. Lund: Lund University; 2007.

[2] Schollwöck U. The density-matrix renormalization group. Rev Mod Phys 2005;77: 259–315.

[3] Hachmann J, Dorando JJ, Avilés M, Chan GKL. The radical character of the acenes: a density matrix renormalization group study. J Chem Phys 2007;127:Art. 134309.

[4] Mazziotti DA. Quantum chemistry without wave functions: two-electron reduced density matrices. Acc Chem Res 2006;39:207–215.

[5] Shepard R, Minkoff M, Brozell SR. Nonlinear wave function expansions: a progress report. Int J Quantum Chem 2007;107:3203–3218.

[6] Parkhill JA, Lawier K, Head-Gordon M. The perfect quadruples model for electron correlation in a valence active space. J Chem Phys 2009;130:Art. 084101.

[7] Chan GKL. Low entanglement wavefunctions. WIREs Comput Mol Sci 2012;2:907–920.

10

THE RAS STATE-INTERACTION METHOD

In a multiconfiguration approach, it becomes important that calculated wave functions can be related to each other even when orbitals differ. One obvious purpose is to compute transition strengths. The problem is that the orbitals simply are not the same if optimized for different states or with different criteria. The same problem is occasionally met also for single-configuration calculations but is more acute in, for example, CASSCF, where the active orbitals can be completely different even for calculations where the *space* spanned by these orbitals are approximately the same. This problem has essentially disappeared with the RASSI (RAS State Interaction) approach.

10.1   THE BIORTHOGONAL TRANSFORMATION

Assume that the two states are described using expansion in Slater determinants, based on orthonormal, but different, orbitals. For simplicity, we describe the principles involved using general spin-orbitals. The CI expansions are

$$\Psi^1 = \sum_\mu C_\mu^X \Phi_\mu^X, \quad \Psi^2 = \sum_\mu C_\mu^Y \Phi_\mu^Y,$$

where Φ_μ^X stands for any type of CSFs built using the spin-orbital basis $\{\psi_p^X\}$. SDs are the natural choice for general spin orbitals, but GUGA or Symmetric Group functions would work as well.

Multiconfigurational Quantum Chemistry, First Edition.
Björn O. Roos, Roland Lindh, Per Åke Malmqvist, Valera Veryazov, and Per-Olof Widmark.
© 2016 John Wiley & Sons, Inc. Published 2016 by John Wiley & Sons, Inc.

As explained in detail in Ref. [1], the "machinery" of any CI program has the capability of computing the effect on the CI coefficients when certain simple operators are applied. This can be used to transform the two sets of CI coefficients to a special "biorthonormal" form. A sequence of special transformation is applied, ending with the result

$$\Psi^1 = \sum_\mu C_\mu^X \Phi_\mu^X = \sum_\mu C_\mu^A \Phi_\mu^A,$$

$$\Psi^2 = \sum_\mu C_\mu^Y \Phi_\mu^Y = \sum_\mu C_\mu^B \Phi_\mu^B.$$

These formulas have been written to emphasize that we have changed the spin-orbital basis sets, using now the new bases A and B, but also with new CI coefficients with the effect that the *wave functions* remain the same. The transformation of orbitals has the form

$$(\psi_1^A, \ldots, \psi_m^A) = (\psi_1^X, \ldots, \psi_m^X) C^{XA},$$

$$(\psi_1^B, \ldots, \psi_m^B) = (\psi_1^Y, \ldots, \psi_m^Y) C^{YB}.$$

This cannot be done in any arbitrary way. First of all, for each of the transformation steps to result in wave functions that are representable as CI expansions with the given class of CSFs, the transformation matrices in the spin-orbital basis, C^{XA} and C^{YB} must be (block-)triangular matrices. Second, the CSF space should be "closed under deexcitation"—a technical term where we note that this is true, for example, for RASSCF wave functions, and for MRCI wave functions with a complete reference.

Finally, to be useful, the transformation matrices should fulfill the requirement

$$C^{YB} C^{XA\dagger} = (S^{XY})^{-1},$$

where S^{XY} contains the overlaps between the two orbital sets, which must be invertible.

For further explanations, we refer to the article [1].

The whole point of this is that the wave functions have been transformed to a pair of CSF bases that forms a *biorthonormal pair*. This allows calculating any transition matrix elements involving the two wave functions essentially as if they were given in a common basis of orthonormal CSFs. Why this is so is easily understood from the so-called *Slater–Löwdin rules* for computing such matrix elements. These are expressed in terms of determinants containing basis function overlaps. If these—as here—have been manipulated into the Kronecker deltas of a biorthonormal pair, the result is formally the same as if a common orthonormal basis had been used.

In the RASSI program, a number of wave functions of the RASSCF type are read, and for each pair, the overlap and the one-body and possibly also two-body density matrices are computed. These are defined as

$$S^{12} = \langle \Psi^1 | \Psi^2 \rangle,$$

$$\gamma_{pq}^{12} = \langle \Psi^1 | \hat{c}_p^\dagger \hat{c}_q | \Psi^2 \rangle,$$

$$\gamma_{pqrs}^{12} = \langle \Psi^1 | \hat{c}_p^\dagger \hat{c}_q^\dagger \hat{c}_s \hat{c}_r | \Psi^2 \rangle.$$

This definition is for some common orthonormal basis. Actually, the density matrices are computed internally using the mixed, nonorthonormal bases (X and Y above) and transformed to such a basis if needed; for matrix elements of properties, they are transformed to the AO basis and contracted with the various property integrals. If Hamiltonian matrix elements are computed, the two-body matrix elements are contracted with two-body AO integrals.

10.2   COMMON ONE-ELECTRON PROPERTIES

The name of this method is from the original use, namely, to compute eigenstates and their transition matrix elements by mixing a set of RASSCF wave functions, much as a nonorthogonal CI mixes configurations; hence "RAS State Interaction." For common one-electron operators such as dipole moment and angular momentum, it is assumed that the relevant integrals over the common AO basis are available. As described above, the sets of property integrals are contracted with the one-body density matrix, and this would seem to be the end of it. However, the states involved may be neither orthonormal nor noninteracting over the Hamiltonian, when they come from separately optimized RASSCF calculations. For special purposes, one may even have prepared a set of wave functions that are not intended to represent energy eigenstates at all, and the matrix elements may be wanted for some other purpose, but usually the calculation is aimed at transitions between such eigenstates. It is then necessary to compute not only the matrix elements of the interesting property as such but also matrix elements of the Hamiltonian and in general also the overlap. This results in a small matrix problem, where the Hamiltonian and overlap matrix elements over the set of RASSCF wave function are entered into matrices H and S, respectively, and the generalized eigenvalue equations are then

$$H v_k = E_k S v_k, \quad k = 1, \dots, N,$$

$$v_k^\dagger S v_l = \delta_{kl}, \quad k, l = 1, \dots, N.$$

Assume the matrix elements over the input states of the operator $\hat{A}$ were computed into the matrix A. The final wave functions and their matrix elements are then

$$\Psi_{\text{rassi}}^k = \sum_{k=1}^{N} v_{i,k} \Psi^i,$$

$$\langle \Psi_{\text{rassi}}^k | \hat{A} | \Psi_{\text{rassi}}^l \rangle = (v)_k^\dagger A v_l.$$

These "RASSI states" cannot in general be represented as a CI expansion but only as the linear combination of the input states. Nevertheless, they are now orthonormal and noninteracting, and therefore the appropriate states, for example, for estimating transition strengths. Their natural orbitals can be computed, and will usually be close to, but not identical to, those of any input state. In fact, the natural occupation numbers will in general deviate from 0 and 2 (assuming now spin-free orbitals) for a larger set of orbitals than the original active space; the active space has become a bit "fuzzy." The "extra" active orbitals are however not good approximations to true natural orbitals; the procedure above is intended only to remove nonorthogonality, and the RASSCF approximation of a limited active space still affects the results.

10.3 WIGNER–ECKART COEFFICIENTS FOR SPIN–ORBIT INTERACTION

In GUGA, wave functions are computed using a spin-independent CSF basis. In our programs, as well as in many other, the CI calculation is done with no explicit assumption on the spin projection M_S, and in fact it can be argued that such a *spin-free method* does not (for $S \neq 0$) represent a wave function at all but rather represents the *space* spanned by the $2S + 1$-dimensional manifold of spin states. This makes it possible to compute matrix elements for spin-dependent operators over the full set of spin states [2].

This approach is not new but has been used a lot by spectroscopists. The group theoretical device is the *Wigner–Eckart* rules, which in this context becomes the use of so-called *reduced matrix elements* and spin-coupling coefficients, such as *Clebsch–Gordan* coefficients or *Wigner's 3j symbols*.

The relevant operator is not necessarily a scalar but could, in principle, be a more complicated entity such as a quadrupole moment with its five components. Such operators are first transformed to a "tensor basis" where the components transform according to a standard irreducible representation of the point group—in this context, this means that the components should be chosen to transform as a "spin tensor." This is a tensor operator $\hat{\mathbf{T}}^S$ with specified spin quantum number S, containing $2S + 1$ components $\hat{T}^S_{M_S}$ with $M_S = -S, -S + 1, \ldots, S$, chosen such that the following commutation relations hold:

$$[\hat{S}_+, \hat{T}^S_{M_S}] = \sqrt{S(S + 1) - M_S(M_S + 1)}\hat{T}^S_{M_S+1},$$

$$[\hat{S}_z, \hat{T}^S_{M_S}] = M_S\hat{T}^S_{M_S},$$

$$[\hat{S}_-, \hat{T}^S_{M_S}] = \sqrt{S(S + 1) - M_S(M_S - 1)}\hat{T}^S_{M_S-1}.$$

For any two states, with spin and spin projections S', M'_S and S'', M''_S, respectively, we get a total number of $(2S' + 1)(2S + 1)(2S'' + 1)$ matrix elements from one single quantity, by the equation

$$\langle \gamma' \ S' \ M'_S | \hat{T}^S_{M_S} | \gamma'' \ S'' \ M''_S \rangle = \langle \gamma' \ S' \| \hat{T}^S \| \gamma'' \ S'' \rangle \times \left(\begin{pmatrix} S' \\ -M'_S \end{pmatrix} \begin{pmatrix} S \\ M_S \end{pmatrix} \begin{pmatrix} S'' \\ M''_S \end{pmatrix} \right)$$

As can be seen, the reduced matrix elements, written with double bars, suffice to compute all the elements, given tabulated 3j-symbols. The symbols γ', γ'' stand for any labels necessary to identify the spin-free states. This is commonly used in formulas for transition strengths involving either atoms or rotational states, and then using angular momentum operators $\hat{L}$ or $\hat{J}$, but is just as convenient for the present purpose.

The way this is done in the RASSI program is to compute not only the usual one-particle transition density matrix but also what we call a "Wigner–Eckart reduced density matrix," which can then be contracted in the usual way with supplied property integrals, and the final result is then combined with numerical factors as needed. Where this is used is primarily with the Atomic Mean-Field Integrals [3] (AMFI). These integrals define an effective one-electron spin-orbit Hamiltonian by the expression

$$\hat{H}^{SO} = \sum_{i} \sum_{\text{Atom } C} \hat{\mathbf{A}}_C \cdot \hat{\mathbf{s}}_i$$

where summation is over electrons i and atoms C. Thus, for each atom, there is a predetermined spin-orbit interaction that contains also the theoretical two-electron terms, where these have been contracted with an average two-electron spin density to produce an effective interaction. The AMFI integrals are computed as matrix elements for (spatial) orbitals of the three components $\hat{A}_x$, $\hat{A}_y$, and $\hat{A}_z$. The actual operator is defined as the scalar product of the vector operator with the individual spin $\hat{\mathbf{s}}_i$ of the electron. The form of a sum over the atoms allows the integrals to be very simply precomputed as any other one-electron property, but this Hamiltonian has proven to be quite accurate when the inner shells are closed.

10.4 UNCONVENTIONAL USAGE OF RASSI

It is possible to use wave functions that do not result from a RASSCF calculation as input to RASSI. One such usage is for obtaining well-defined diabatic states, which is useful, in the vicinity of curve crossing or conical intersections, since these states give stable matrix elements for fitting, for example, the diabatic potentials and electrostatic couplings. In the simplest cases, one can select some property, for example, dipole moment, and devise the transformation to diabatic states by hand or by computer such that this property keeps continuous and stable—a small set of wave functions can be used as basis functions, and the matrix of one dipole moment component is diagonalized in this basis.

This cannot serve as a general recipe, however. A more elaborate approach has been to define diabatic states as linear combinations of predefined, specially prepared set of wave functions. The MO- and CI-coefficients of these states stay constant, while the structure, and therefore the atomic basis, varies which provides a stable and smoothly varying set of reference states. They will neither stay orthonormal nor normalized but smoothly depend on the structure. By computing overlaps using RASSI, one obtains a general and flexible scheme for defining quasi-diabatic linear combinations of the energy eigenstates. At the same time, this eliminates the simpler, but often irritating, problem of keeping track of relative phases of the orbitals and states, which may undergo erratic sign shifts during the course of a calculation.

10.5   FURTHER READING

- A.J. Ceulemans, "Group Theory Applied to Chemistry", Springer-Verlag, Dordrecht, 2013.

10.6   REFERENCES

[1] Olsen J, Godefroid MR, Jönsson P, Malmqvist PÅ, Froese-Fischer C. Transition probability calculations for atoms using non-orthogonal orbitals. Phys Rev E 1995;52:4499–4508.

[2] Malmqvist PÅ, Roos BO, Schimmelpfennig B. The restricted active space (RAS) state interaction approach with spin-orbit coupling. Chem Phys Lett 2002;357:230–240.

[3] Schimmelpfennig B. AMFI, An Atomic Mean-Field Spin-Orbit Integral Program. Sweden: Stockholm University; 1996.

11

THE MULTIREFERENCE CI METHOD

For any given one-particle basis set, the most accurate treatment within the algebraic approximation method is termed *Full CI*, which implies solving the eigenvalues problem using all the CSFs (or SDs) that can be constructed using the basis set at hand. For any but the very smallest basis sets, this is impossible.

11.1   SINGLE-REFERENCE CI. NONEXTENSIVITY

Assuming n_o spatial orbitals, n_α and n_β electrons with spin up and spin down, respectively, and without any useful point group symmetry, the number of SDs we can form is

$$N_{\mathrm{SD}} = \binom{n_o}{n_\alpha} \binom{n_o}{n_\beta}. \tag{11.1}$$

Usually, a number of core orbital can and should be *frozen*, that is, left uncorrelated, and always doubly occupied. Such orbitals can usually be optimized by a CASSCF procedure (say), and used without further optimization. For example, a basis set of so-called "valence double-zeta plus polarization," or VDZP, quality for ammonia will contain 28 basis functions apart from the frozen 1s core, and gives rise to 419 million SDs. For such large numbers, the notation "419 MDets" is convenient. Using instead singlet-coupled CSFs, we similarly get 97 MCSFs. Such calculations are quite feasible. They are also quite meaningless, except that they are very useful

Multiconfigurational Quantum Chemistry, First Edition.
Björn O. Roos, Roland Lindh, Per Åke Malmqvist, Valera Veryazov, and Per-Olof Widmark.
© 2016 John Wiley & Sons, Inc. Published 2016 by John Wiley & Sons, Inc.

as yielding benchmarks against which more approximate methods can be compared. They are meaningless if the aim is an accurate description of ammonia and its properties, since much higher accuracy could be obtained with a large basis set and a less ambitious treatment of the correlation.

This calculation can and has been performed. But consider a similar calculation for the very small molecule C_2H_5OH with 71 valence orbitals and 18 valence electrons; this gives 5.5×10^{21} SDs or 0.63×10^{21} CSFs. It is obvious that the Full CI will usually remain a formal standard that cannot be reached computationally. In an infinite basis set, the Full CI is called *Complete CI*.

One common way of performing a reasonable-sized calculation on larger systems is by truncating the expansion. Starting from a single determinant, "singles," "doubles," etc. refer to other SDs formed by replacing one, two, etc. occupied spin-orbitals with orbitals that are unoccupied in the reference determinant. This is a simple scheme that can work quite well. The resulting SDs are also called "singly excited," relative to the reference, which usually (but not always) is the ground state Hartree–Fock state. The drawback is a loss of extensivity.

The total energy, such as some other properties of real molecular systems, is *extensive*. This is a term from thermodynamics and means that if several subsystems are isolated from each other, one should be able to get the total energy from computing the energy of each subsystem and then add them. Within quantum chemistry, a method is called *size extensive* if it has this property. However, as soon as a CI expansion is truncated to some particular excitation level, we lose extensivity.

This can be illustrated as follows: Assume a system consisting of N identical subsystems, which are isolated from each other. Each subsystem contains two electrons and is accurately represented by a wave function Φ_k, where $k \in \{1, ..., N\}$. We assume further that each such wave function is accurately represented by two configurations,

$$\Phi_k = c|g_{k\alpha}g_{k\beta}\rangle - s|u_{k\alpha}u_{k\beta}\rangle,$$

where $c^2 + s^2 = 1$, for a normalized wave function.

This is a perfect pairing wave function and could be a first ansatz for a system with a lot of correlated bonds, assumed not to interact with each other, or a number of correlated H_2 molecules separated by large distances.

Assuming the reference SD to be the determinant function

$$|0\rangle = |g_{1\alpha}g_{1\beta}g_{2\alpha}g_{2\beta} \cdots g_{N\alpha}g_{N\beta}\rangle$$

we find the total wave function to be composed as

$$C_0|0\rangle + C_D|D\rangle + C_Q|Q\rangle + ...,$$

where

$$C_0 = c^N, \quad C_D = -\sqrt{N}c^{N-1}s, \quad C_Q = \sqrt{\frac{N(N-1)}{2}}c^{N-2}s^2, \ ...$$

Performing the calculation using SDCI, we would get a perfectly accurate result for any of the subsystems, but for a composite system, the wave function is not contained in the expansion space—it contains only double excitations, not the quartic and higher ones. If the exact correlation energy for N subsystems is $\epsilon_c^{FCI}(N)$, and the SDCI recovers $\epsilon_c^{SD}(N)$, we get

$$\epsilon_c^{FCI}(N) = N\epsilon_c^{FCI}(1) = N\epsilon_c^{SD}(1),$$

$$\epsilon_c^{SD}(N) = \left(\frac{2}{B + \sqrt{B^2 + 4NA}}\right)\epsilon_c^{FCI}(N),$$

where $A = (s/c)^2, B = 1 - A$.

The first line shows that the Full CI result scales correctly with N; the second line shows that, for $N > 1$, the SDCI result is only some fraction of the correct Full CI result. Inserting reasonable values, $\epsilon_c^{FCI}(1) \approx 10\,kcal/mol$ and $s/c \approx 0.1$, shows that the extensivity error reaches above 1 kcal/mol for $N = 4$ subsystems, 10 kcal/mol for $N = 11$, and then rises fast.

11.2 MULTIREFERENCE CI

Occasionally, CI has been truncated at higher excitation levels, but this is very expensive and seldom worthwhile. One may think of the higher excitation levels as being needed only for some selected subset of orbitals; this has led to the *Multireference CI*, MRCI, which is almost always taken to the MRCISD (or SDMRCI) level. Excitations or orbital replacements are then relative to a selected subset of determinants. A replacement is single if it is obtained from at least one of the reference SDs by a single replacement, and so on. Even if the MRCI is not a size-extensive method, the error is minimized by applying any one of a number of corrections, as will be seen shortly.

The first implementation of a general MRCI program was done by Buenker and Peyerimhoff [1, 2]. They, and others, experimented with different approaches to the problem of selecting references, as well as selecting individual CSFs to be included in the CI. Rapid development of computers and of linear algebra software allow nowadays very fast CI routines provided that the CI coefficients can be arranged in systematic ways, which prevent individual selection.

From the numbers given above, it is evident that the use of spin-coupled CSFs gives significantly smaller CI problems to be solved. This is not a decisive factor, since using SDs, on the other hand, gives simpler computer codes and—most important—are better suited to use the advantages of optimized linear algebra libraries. An important point is that spin-coupled CSFs automatically give control of the spin of the wave function and prevent so-called *spin contamination*. Many ways have been used for efficient spin-adaption, with names such as Rumer and Serber functions, but the dominating two schemes used today are the Graphical Unitary Group approach [3, 4] (GUGA), which generates irreducible representations of the group of unitary orbital transformations, and the Symmetric Group approach, which uses Young Tableaux [5, 6] to generate CSFs that are irreducible representations of the symmetric group (i.e., permutation symmetry of spatial orbitals).

Another MRCI code was developed by Siegbahn and Roos [7, 8], introducing the powerful method of *Direct CI*. This implies that a Hamiltonian matrix is never formed or explicitly stored. Instead, the CI Hamiltonian is represented by the MO integrals and by a set of *coupling coefficients*, such that for any CI vector v, a subroutine call can quickly generate the vector σ, by the definitions

$$\sigma = Hv.$$

The notation is standard enough that one often speaks of a "sigma routine" in this context. The GUGA approach was instrumental for this development. This is the key to using Davidson's method for computing a few eigenvalues and corresponding eigenvectors of very large matrices [9], which are never constructed or stored as such. This basic method has been further developed into very efficient codes used today, which allow calculations with on the order of 10^{11} SDs [10, 11].

The size extensivity corrections mentioned earlier are necessary for any except the very smallest calculations (except of course for FCI). The most well-known is the Davidson correction [12] or similar, originally intended for plain single-reference CI, but which has been generalized for MRCI. Such corrections are also called "corrections for unlinked quadruples" and are generally recognized by appending "+Q" to the method name. There are also corrected schemes relying on a modification of the CI equations. A very good such scheme, which can be extended to multireference calculations, is the *Averaged Coupled Pair Functional* (ACPF) method originally by Gdanitz and Ahlrichs [13], and some later variants.

A versatile modern implementation of MRCI, using the GUGA approach, is found in the COLUMBUS code [14, 15].

11.3 FURTHER READING

We can recommend the following articles:

- T.S. Chwee, A.B. Szilva, R. Lindh, and E.A. Carter, "Linear scaling multireference singles and doubles configuration interaction", J Chem Phys 128, Art. 224106 (2008).
- I. Shavitt, "The history and evolution of configuration interaction", Mol Phys 94, 3 (1998).

11.4 REFERENCES

[1] Peyerimhoff SD, Buenker RJ. Comparison of various CI treatments for the description of potential curves for the lowest three states of O_2. Chem Phys Lett 1972;16: 235–243.

[2] Buenker RJ, Peyerimhoff SD. Individualized configuration selection in CI calculations with subsequent energy extrapolation. Theor Chim Acta 1974;35: 33–38.

[3] Paldus J. Group theoretical approach to the configuration interaction and perturbation theory calculations of atomic and molecular systems. J Chem Phys 1974;61: 5321–5330.

[4] Shavitt I. The graphical unitary group approach and its application to direct configuration interaction calculations. In: Hinze J, editor. The Unitary Group for the Evaluation of Electronic Energy Matrix Elements. Berlin: Springer-Verlag; 1981. p. 51–99.

[5] Yamanouchi T. On the calculation of atomic energy levels. Proc Phys Math Soc Jpn 1935;17: 274–288.

[6] Kotani M, Ohno K, Kayama K. Encyclopedia of physics. In: Flügge S, editor. Encyclopedia of Physics. vol. 37/2. Berlin: Springer-Verlag; 1961. p. 118–142.

[7] Roos BO. A new method for large-scale CI calculations. Chem Phys Lett 1972;15: 153–159.

[8] Roos BO, Siegbahn PEM. The direct configuration interaction method from molecular integrals. In: Schaefer HF, editor. Modern Theoretical Chemistry. vol. 3. New York: Plenum Press; 1977. p. 277–318.

[9] Davidson ER. The iterative calculation of a few of the lowest eigenvalues and corresponding eigenvectors of large real symmetric matrices. J Chem Phys 1975;17: 87–94.

[10] Olsen J, Jørgensen P, Simons J. Passing the one-billion limit in full configuration-interaction (FCI) calculations. Chem Phys Lett 1990;169: 463–472.

[11] Mitrushenkov AO. Passing the several billions limit in FCI calculations on a mini-computer. Chem Phys Lett 1994;217: 559–565.

[12] Langhoff SR, Davidson ER. Configuration interaction calculations on the nitrogen molecule. Int J Quantum Chem 1974;8: 61–72.

[13] Gdanitz RJ, Ahlrichs R. The averaged coupled-pair functional (ACPF): a size-extensive modification of MR CI(SD). Chem Phys Lett 1988;143: 413–420.

[14] Yabushita S, Zhang Z, Pitzer RM. Spin-orbit configuration interaction using the graphical unitary group approach and relativistic core potential and spin-orbit operators. J Phys Chem A 1999;103: 5791–5800.

[15] Lischka H, Müller T, Szalay PG, Shavitt I, Pitzer RM, Shepard R. Columbus - a program system for advanced multireference theory calculations. WIREs Comput Mol Sci 2011;1: 191–199.

12

MULTICONFIGURATIONAL REFERENCE PERTURBATION THEORY

12.1 CASPT2 THEORY

12.1.1 Introduction

For closed-shell or high-spin open-shell states, if it can be reasonably well approximated with a single Slater determinant, there are efficient methods for including dynamic correlation effects, such as perturbation theory and coupled cluster. There are variants of multiconfigurational coupled cluster, but these seem to address computations with only very few configurations.

A popular method is MP2 – Møller–Plessett perturbation theory through second order. Adding or including dynamic correlation effects in a similar way for multiconfigurational root states is not easy. However, in the case of CASSCF wave functions, a Møller–Plessett-like approach exists. CASPT2 (CAS Perturbation Theory through second order) [1, 2] is a method for computing a second-order dynamic correlation contribution "on top" of CASSCF. In principle, this is conventional Rayleigh–Schrödinger perturbation theory, for a single electronic state function. However, the unperturbed state (often called the root state, or root function) is not a single-determinant state but a CASSCF state, typically including a few hundred thousand up to a few million configuration state functions. Therefore, any type of state (radical, charged, excited, etc.) can be used, as long as it is adequately described by a CASSCF wave function. The Multistate CASPT2 can also handle

Multiconfigurational Quantum Chemistry, First Edition.
Björn O. Roos, Roland Lindh, Per Åke Malmqvist, Valera Veryazov, and Per-Olof Widmark.
© 2016 John Wiley & Sons, Inc. Published 2016 by John Wiley & Sons, Inc.

simultaneously several root states. RASSCF, rather than CASSCF, root functions can be used, but some double intra-active excitations are then neglected.

Other perturbative methods exist, for example, the so-called NEVPT2 method (N-Electron Valence State Perturbation Theory) by Angeli et al. [3] and multireference QDPT (Quasi-Degenerate Perturbation Theory) methods by Nakano; see, for example, Ref. [4]. For an overview, see Ref. [5]. Evaluation and comparison to other methods can be found, for example, in articles by Ghosh et al. [6] and Schreiber et al. [7].

The combination CASSCF/CASPT2 has turned out to be very successful for a large variety of problems. While CASSCF alone can handle nondynamic correlation with a fairly large number of active orbitals, it lacks dynamic correlation, which is then provided by the perturbative calculation. The accuracy that can be reached is then dependent on the basis set, and on the ability of the CASSCF to account for nondynamic correlation. In a study of atoms and diatoms, Ghigo et al. [8] found that the errors in relative energies (dissociation and excitation energies) were generally of the order of 0.1 eV, or about 2 kcal/mol, allowing reliable assignments of the features in electronic spectra.

In discussing this method, we may compare to the (usually unfeasible) full CI, using the entire set of configuration functions that can be formed from the MO basis. The "root states" are assumed to have fully converged CI expansions. The orbitals are typically not optimized for each state but are obtained by a State-Averaged CASSCF or RASSCF. This is partly for convenience, but also because it is generally difficult to optimize orbitals for individual excited states. For very small systems (atoms and small molecules) and basis sets with DZP or TZP quality such full CI calculations have been done, and show CASPT2 to yield precision of essentially MP2 quality when the root state is close to a closed-shell SCF state, but retaining this precision when the MP2 fails due to nondynamic correlation that is satisfactorily treated by CASSCF. Note that MP2 is a limiting case when the CASSCF root state is in fact a closed-shell Hartree–Fock state.

12.1.2   Quasi-degenerate Rayleigh–Schrödinger perturbation theory

Let $\mathbf{H}$ be a given Hermitian $N \times N$ matrix, for which we wish to compute an approximation to some class of eigenvalues, such as the single lowest or a few of the lowest eigenvalues, or selected in some different way. We say that the corresponding eigenvectors span the selected subspace of vectors, also called the target subspace, and it has dimension n.

Suppose that we can approximate the matrix with another Hermitian $\mathbf{H}^{(0)}$, with the properties that all its eigenvalues and eigenvectors are known, and define a new parametrized matrix $\mathbf{H}(t)$ such that

$$\mathbf{H}(t) = (1 - t)\mathbf{H}^{(0)} + t\mathbf{H},$$

$$\mathbf{H}(0) = \mathbf{H}^{(0)} = \mathbf{U}^{(0)}\mathbf{D}^{(0)}\mathbf{U}^{\dagger(0)}.$$

The eigenvectors of $\mathbf{H}^{(0)}$ have been arranged as columns of $\mathbf{U}^{(0)}$. The matrix $\mathbf{D}^{(0)}$ is a diagonal matrix, and its diagonal elements are the eigenvalues of $\mathbf{H}^{(0)}$. Clearly, the second row of the equation above simply expresses the eigenvalue equations in matrix form. From now on, we can use these columns as an orthonormal basis. Consider this done, and by ordering them such that the first n approximately span the target subspace, the equations can be displayed in the form of (2×2) matrices, whose elements are submatrices. The space spanned by the first n eigenvectors is called the model space, and is often indicated by using the index P, while the complementary space with dimension $N - n$ has index Q.

$$\mathbf{H}(t) = \begin{pmatrix} \mathbf{H}_{PP}^{(0)} & \mathbf{0} \\ \mathbf{0} & \mathbf{H}_{QQ}^{(0)} \end{pmatrix} + t \begin{pmatrix} \mathbf{H}_{PP}^{(1)} & \mathbf{H}_{PQ}^{(1)} \\ \mathbf{H}_{QP}^{(1)} & \mathbf{H}_{QQ}^{(1)} \end{pmatrix}.$$

We are interested only in the first n eigenvalues and eigenvectors, which can be obtained as the solution of the equations

$$\mathbf{H}_{PP}^{(0)} + t\mathbf{H}_{PP}^{(1)} + t\mathbf{H}_{PQ}^{(1)}\mathbf{X}(t) = \mathbf{H}_{\text{eff}}(t),$$

$$t\mathbf{H}_{QP}^{(1)} + \left(\mathbf{H}_{QQ}^{(0)} + t\mathbf{H}_{QQ}^{(1)}\right)\mathbf{X}(t) = \mathbf{X}(t)\mathbf{H}_{\text{eff}}(t).$$

12.1.3   The first-order interacting space

Consider one of the root states, with wave function $\Psi^{(0)}$, assumed to be close to an exact wave function Ψ. A measure of the error is given by the residual wave function,

$$\rho = \left(\hat{H} - E_0\right)\Psi^{(0)},$$

where $\hat{H}$ is the full CI wave function, $\|\rho\|^2$ is minimized by choosing E_0 to be the variational CASCI energy, and the residual is then orthogonal to all the CASCI space.

It should be pointed out that the deeper core orbitals are usually excluded from the correlation treatment. This is not just a simplification but is always done for wave function-based correlation calculations, unless very special basis sets or methods are used; else, the large dynamic correlation contribution from the core orbitals would give a very large basis set superposition error (see Chapter 6). Neglecting this core correlation gives instead an almost constant error. The excluded orbitals are termed "frozen," in the context of correlation.

The residual vector lies completely within the so-called first-order interacting space, which is that generated by all terms in the Hamiltonian and orthogonal to the CASCI space. Moreover, each such term is (in second quantization) generated by a one- or two-electron excitation, with at least one nonactive orbital index. For a Multireference SDCI calculation, this space would be spanned by all Slater Determinants (SDs), that are not in the reference, but differ by at most a two-electron excitation from at least one reference SD. This space can be huge, even for rather small calculations. One of the strong points of perturbation theory is that the first-order interacting space can be kept smaller.

In the conventional *Møller–Plessett* perturbation theory, the wave function is a single determinant, and the Hamiltonian is approximated by assuming that each wave function term generated (as above) has a well-defined energy that differs from the root state by a sum or difference of orbital energies. The first-order interacting space of SDs is here identically the same as the space generated by one- and two-electron excitations. The first-order wave function can be used to compute the perturbation energy of up to first, second, or even third order; the most common choice is MP2, the energy through second order.

In CASPT2, a similar perturbation energy is computed, and it is easily seen to become equivalent to MP2 in the limit that (i) the root function is a single-determinant function, (ii) its coupling with excitation terms is weak enough, and (iii) the approximation to the Hamiltonian is by using orbital energies. However, the wave function terms that span the first-order interaction space are in general neither orthonormal, nor noninteracting over the approximate Hamiltonian. They are, on the other hand, very much fewer than the corresponding space of SDs.

12.1.4 Multiconfigurational root states

CASPT2 uses a CASSCF wave function as the unperturbed wave function, but it has also been extended to use a more economic RASSCF reference.

While perturbation techniques are well understood generally, the use of general multiconfigurational root functions demands several technically motivated differences from those with a single-configuration root function [5].

Using Rayleigh–Schrödinger perturbation theory, the Hamiltonian operator is divided up in two parts, $\hat{H} = \hat{H}_0 + \hat{H}_1$, where $\hat{H}_0$ has a known eigenfunction. The parametrized Schrödinger equation

$$\left(\hat{H}_0 + \lambda\hat{H}_1\right)\Psi_\lambda = E_\lambda\Psi_\lambda$$

allows a solution in the form of a Taylor expansion,

$$\Psi_\lambda = \sum_{n=0}^{\infty} \Psi_n \lambda^n \qquad (12.1)$$

$$E_\lambda = \sum_{n=0}^{\infty} E_n \lambda^n \qquad (12.2)$$

with some side condition on of Ψ_λ, usually the intermediate normalization,

$$\langle \Psi_\lambda | \Psi^{(0)} \rangle = 1.$$

The solution as a truncated power series through order n in λ is obtained as a recursive solution of the equations

$$
\left.\begin{array}{rcl}
E_{2n} & = & \langle\Psi_n|\hat{H}_0|\Psi_n\rangle \\
E_{2n+1} & = & \langle\Psi_n|\hat{H}_1|\Psi_n\rangle \\
(\hat{H}_0 - E_0)\Psi_{n+1} & = & -\left(\hat{H}_1\Psi_n - \sum_{k=1}^{n-1} E_k\Psi_{n-k}\right)
\end{array}\right\}, \tag{12.3}
$$

where the side relation $\langle\Psi_n|\Psi^{(0)}\rangle = 0$, for $n = 1, 2, \cdots$, is also assumed.

Obviously, the zeroth-order Hamiltonian cannot be chosen freely, if one has at hand a general zeroth-order wave function, or root function, since this must be an eigenfunction of $\hat{H}_0$. This is no problem in a conventional scheme, like Møller–Plessett, which use an $\hat{H}_0$ in the form of a one-electron operator and a $\Psi^{(0)}$ which is a single-determinant, or a single-configuration state function built from orbitals that are eigenfunctions to this one-electron Hamiltonian. As a simple example, consider a wave function with four electrons in three orbitals, singlet, which can have the form

$$
|\Psi^{(0)}\rangle = c|\psi^{(1)}\overline{\psi}_1\psi_2\overline{\psi}_2\rangle - s|\psi^{(1)}\overline{\psi}_1\psi_3\overline{\psi}_3\rangle = c|1\overline{1}2\overline{2}\rangle - s|1\overline{1}3\overline{3}\rangle,
$$

where we use an overbar to indicate β spin, and write the determinants in an abbreviated fashion, with orbital 1 inactive, orbitals 2 and 3 active. If we assume that orbital 3 acts as correlating orbital, so that $c \approx 1$ and s is small, then in general orbital 3 is far from being an approximate SCF orbital, it is close in shape to orbital 2 but with an additional node surface to account for orthonormality, and the orbital energy (reasonably defined) of orbital 3 will be high – probably higher than the lowest virtual energies if a decent basis set is used.

Then we have the following problems. While a single determinant would allow every CSF to be one-to-one related with a corresponding excitation operator, like a double excitation from inactive to virtual, this is not the case here. First, consider, for example, the two excitation operators $\hat{E}_{t2t2}$ and $\hat{E}_{t3t3}$ applied to $\Psi^{(0)}$: these produce the same result, $|1\overline{1}\,t\overline{t}>$. Furthermore, in order to solve for the first-order wave function, the perturbation equations should be sparse, ideally diagonal. This is the responsibility of the approximative Hamiltonian. Ideally, the wave function terms produced by each excitation should be orthogonal and noninteracting (over the approximative Hamiltonian). This is no longer possible if $\hat{H}_0$ is a Fock-type operator, that is, $\hat{H}_0 = \sum \epsilon_p \hat{a}_p^\dagger \hat{a}_p$ where $\hat{a}_p^\dagger$ and $\hat{a}_p$ are creation and annihilation operators for some suitably chosen orbitals. In fact, such an operator would even fail to have the simple two-configuration root function as an eigenfunction, unless of course orbitals ψ_2 and ψ_3 had the same orbital energy. This would not be a good choice when ψ_3 is a correlating orbital.

Finally, consider excitations from the inactive to the active space.

$$
\hat{E}_{21}\Psi^{(0)} = -s\frac{1}{\sqrt{2}}(|1\overline{2}3\overline{3}\rangle - |\overline{1}23\overline{3}\rangle) \text{ and } \hat{E}_{31}\Psi^{(0)} = c\frac{1}{\sqrt{2}}(|12\overline{2}\overline{3}\rangle - |\overline{1}2\overline{2}3\rangle).
$$

In the first case, we might expect to produce a wave function term with energy (relative to the root state) of about $\epsilon_2 - \epsilon_1$, but instead get a highly excited state because the excitation, apart from the expected single replacement, also *selects* terms with the correlating orbital doubly occupied. In the second case, it works as one would expect. There is no way to choose orbital energies such that energy denominators become the usual orbital energy sums/differences.

In the case of CASPT2, these problems are avoided by formulating the equations as matrix equations, where the matrix elements of the approximate Hamiltonian are explicitly computed over the wave function terms arising from each excitation operator, and by deliberately excluding from $\hat{H}_0$ any coupling that transfers electrons between inactive, active, or virtual subspaces. This matrix has in principle an order equal to the number of excitations, but it can be factorized, and the perturbation equations are solved by an efficient preconditioned conjugate method.

The problems are primarily caused by sticking to a model where individual nondynamically correlated state functions are used as root functions, making in effect the excited wave function terms "precontracted" by the CASCI. The alternative would be to use the dynamic correlation first, to obtain a "dressing" of the internal CAS Hamiltonian, but this procedure has problems of its own, and we prefer to start with CASSCF states. Such a solution is used in the NEVPT2 method. This is a quasi-degenerate perturbation scheme using the so-called CAS/A $\hat{H}_0$ of Ref. [9], and several of the complexities here are thereby avoided, which leads to a very efficient perturbation scheme. However, we only describe the CASPT2 in some detail here.

12.1.5   The CASPT2 equations

Given a CASSCF wave function $\Psi^{(0)}$, the Rayleigh–Schrödinger equation for the first-order wave function $\Psi^{(1)}$ has the conventional form, for example,

$$(\hat{H}_0 - E_0)\Psi^{(1)} = -(\hat{H} - E_0)\Psi^{(0)} = \Psi^{(\text{RHS})}.$$

The equation is soluble if $\Psi^{(0)}$, the root function, is an eigenfunction to the approximate Hamiltonian operator $\hat{H}_0$ with eigenvalue E_0. The solution is unique if it is restricted to be orthogonal to $\Psi^{(0)}$, and if E_0 is distinct from all other eigenvalues.

Let orbital subsets be indicated simply by the letters used for indices: we use p, q, r, and s to stand for any arbitrary orbital; while i, j, k, and l are reserved for inactive; t, u, v, and x for active; and a, b, c, and d stand for virtual orbitals. Usually, we wish to have a spin-independent formalism. It is then conventional to replace the annihilation and creation operators by so-called spin-free or spin-summed excitation operators:

$$\hat{E}_{pq} = \sum_\sigma \hat{a}_{p\sigma}^\dagger \hat{a}_{q\sigma}, \quad \hat{E}_{pqrs} = \sum_{\sigma,\sigma'} \hat{a}_{p\sigma}^\dagger \hat{a}_{r\sigma'}^\dagger \hat{a}_{s\sigma'} \hat{a}_{q\sigma}.$$

The CI Hamiltonian can then be written as

$$\hat{H} = \sum_{pq} h_{pq} \hat{E}_{pq} + \frac{1}{2} \sum_{pqrs} (pq, rs) \hat{E}_{pqrs}.$$

TABLE 12.1   The Excitation Types Used in CASPT2

Internal	Semi-Internal	External
$\hat{E}_{tiuv}$	$\hat{E}_{atuv}$	$\hat{E}_{atbu}$
$\hat{E}_{tiuj}$	$\hat{E}_{aitu}$ or $\hat{E}_{tiau}$	$\hat{E}_{aibt}$
	$\hat{E}_{tiaj}$	$\hat{E}_{aibj}$

For a CASSCF-type wave function, with converged CI coefficients and energy E_0,

$$\left(\hat{H} - E_0\right)\Psi^{(0)} = \rho,$$

where ρ, the residual, can be written exactly in terms of an expansion

$$\rho = \sum_{KP} R_P^{(K)} \hat{X}_P^{(K)} \Psi^{(0)}, \tag{12.4}$$

$$\hat{X}_P^{(K)} \in \left\{ \hat{E}_{tuvj}, \hat{E}_{tiuj}, \cdots, \hat{E}_{bjai} \right\}, \tag{12.5}$$

where K is an index denoting excitation type and P is index of an individual excitation of that type. It turns out that the interacting space is spanned by eight different types of operators, with the suggestive abbreviations "TUVX," ... ,"BJAI" [1, 2]. These are two-electron operators $\hat{E}_{pqrs}$ where indices q, s – the annihilator indices – denote orbitals that are occupied in the CASSCF, and the creator indices p, r are either active or virtual. Also, since the CI coefficients are assumed to be converged, excitation operators with four active indices are not needed: ρ will contain no components that were included in the CI. Finally, it is noted that one-electron excitations are redundant: unless the active space is empty, they are linearly dependent on the two-electron excitations (excepting the extreme case with no active space). This general fact does not in any way depend on orbital optimization. The excitation types are listed in Table 12.1. These excitations spans all one- and two-electron terms present in the Hamiltonian, and is thus called the "first-order interacting" space. The so-called semi-internal excitations would not be present in a single-configuration calculation but are usually quite important in the multiconfigurational case.

If the $\hat{H}_0$ operator could be chosen as a simple one-electron operator, which furthermore does not couple the inactive, active, and virtual orbital spaces, then this could be trivially diagonalized, resulting in quasi-canonical orbitals. The CASSCF root function is easily reexpressed using these orbitals, and the calculation would be similar to the well-known Møller–Plessett perturbation theory, for which

$$\hat{H}_0 = \sum_p \epsilon_p \hat{E}_{pp}.$$

This will not do in the case of a multiconfiguration perturbation theory. Relaxing the demands on the $\hat{H}_0$ operator somewhat, we can at least request that the first-order wave function lies in the first-order interacting space and can be written as

TABLE 12.2   Comparison Between CASPT2 and MP2

Calculation	Type	a' Orbitals	a'' Orbitals	Comment
SCF	Occupied	57	41	
	Virtual	229	207	
	Total	286	248	$= 534$ basis functions
MP2	Frozen	21	13	
	Correlated	36	28	195×10^6 parameters
CASSCF	Inactive	52	39	
	Active	9	5	1.4×10^6 SDs
CASPT2	Frozen	21	13	
	Inactive	36	28	
	Active	9	5	247×10^6 parameters

$$\Psi^{(1)} = \sum_{KP} t_P^{(K)} \hat{X}_P^{(K)} \Psi^{(0)} \tag{12.6}$$

with excitation operators as defined above.

This gives a representation in terms of orbital excitation amplitudes, not many more than is used in a single-determinant-based CISD, MP2, or CCSD calculation. The perturbation equations are projected on the space of interacting wave functions $\hat{X}_P^{(K)} \Psi^{(0)}$. This is an internally contracted space since each such function comprises a large set of SDs but with relative importance predetermined by CI coefficients in the $\Psi^{(0)}$ wave function. As an example, assume a singlet CASSCF wave function such as the one for the chloroiron corrole complex, which is reported in the application section in this book. One of the calculations had 14 active electrons in 14 orbitals, with 534 orbitals altogether, with resulting sizes as given in Table 12.2.

The parameters used in the perturbation methods are called "amplitudes" and are the coefficients that multiply the excitation operators. For MP2, the number of such parameters is on the order of the number of pairs of occupied orbitals, times the number of pairs of virtual orbitals, times two: the pairs may be either singlet or triplet coupled. For CASPT2, one must also include some more excitation amplitudes, since electrons can be excited into as well as excited from the active orbitals.

The use of the excitation amplitudes in the CASPT2, which has the effect of using an internally contracted n-electron basis, results in a number of parameters not much larger than for a similar Møller–Plesset (MP2) calculation. In contrast, if the calculations would be done as a so-called MR-SDCI calculation, using the CASSCF wave function space as a reference, one would need the entire interacting space of Slater Determinant (SD), containing about 1.4×10^{16} SDs. This number is too large to allow us to use uncontracted SDs. On the other hand, requiring the solution to lie in the interacting space, but in the contracted form, is expected to be a good approximation to the corresponding solution in the interacting space of SDs. The equations are then projected on the same set of contracted wave functions, and we get equation systems of the type

$$\sum_{LQ} A_{PQ}^{(KL)} t_Q^{(L)} = u_P^{(K)}$$

with the definitions

$$A_{PQ}^{(KL)} = \left\langle \Psi^{(0)} \left| \left(\hat{X}_P^{(K)} \right)^{\dagger} \left(\hat{H}_0 - E_0 \right) \hat{X}_Q^{(L)} \right| \Psi^{(0)} \right\rangle$$

$$u_P^{(K)} = - \left\langle \Psi^{(0)} \left| \left(\hat{X}_P^{(K)} \right)^{\dagger} \hat{H} \right| \Psi^{(0)} \right\rangle, \tag{12.7}$$

This equation system will be called the CASPT2 equations. Figure 12.1 shows in gray the non-zero elements of the Full CI Hamiltonian matrix.

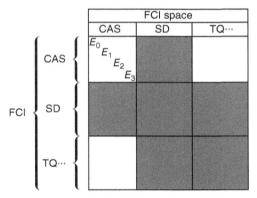

Figure 12.1 The Full CI Hamiltonian matrix structured using the CASPT2 excitation operators. Gray areas: Nonzero matrix elements. (Not drawn to scale).

Where K and L are indices for excitation type, P and Q are indices of individual excitations. The excitation operators, $\hat{X}$, have so far been just the primitive excitation operators. Trivial linear dependencies due to permutation symmetries are obviously to be removed, but there are additional dependencies or near dependencies that can be removed by using appropriate linear combinations of the primitive operators. Those excitation operators that try to remove electrons from very weakly occupied active orbitals, and to move them to strongly occupied ones, result in very small values of these matrix elements. This has the consequence that, even with scaling of the terms to normalize the diagonal values of the matrix A, the condition number of the resulting equation systems is often very large. In fact, a toy example earlier showed that there may even be linear dependencies. Only one way has been found so far of solving the linear equation systems, and this is by symmetrical orthonormalization of the wave function terms used to represent the perturbation problem. All inactive or virtual indices give Kronecker delta factors, but this still leaves us with the necessity to make a full diagonalization of two matrices of the order n^3 if n is the number of active orbitals. For 12 active orbitals or less, this a fairly trivial effort, given today's efficient linear algebra subroutine libraries, but with more than 20, it becomes a considerable undertaking. Nevertheless, this is a viable route.

The actual $\hat{X}^K$ operators have then definitions that depend on the type K, as exemplified by the "VJTU" case:

$$\hat{X}_{Ai}^{\text{VJTU}} = \sum_{tuv} T_{A,tuv}^{\text{VJTU}} \hat{E}_{tiuv},$$

where the transformation matrices—$T_{A,tuv}^{\text{VJTU}}$ in this example—have been determined such that

$$\langle \hat{X}_{Ai}^{\text{VJTU}}\Psi^{(0)}|\hat{X}_{Bj}^{\text{VJTU}}\Psi^{(0)}\rangle = \delta_{ij}\delta_{AB},$$

$$\text{and} \quad \langle \hat{X}_{Ai}^{\text{VJTU}}\Psi^{(0)}|\hat{H}_0|\hat{X}_{Bj}^{\text{VJTU}}\Psi^{(0)}\rangle = \delta_{ij}\delta_{AB}(\Delta_A^{\text{VJTU}} + \epsilon_A^{\text{VJTU}}).$$

The expressions factorize into Kronecker delta factors for the inactive orbitals (and for virtual indices, in other cases) and dense matrices involving the active orbitals. When these have been diagonalized, there is also simply Kronecker deltas for the specific linear combinations of the active orbitals.

Since the equation matrix is now diagonal, the solution after this step would be direct and trivial. However, there is one final complication: the solution as sketched above requires that $\hat{H}_0$ does not couple different types of excitations—it has been assumed that there results one equation system for each individual index K. This is not the case.

As said before, we would like to have an $\hat{H}_0$ of comparable simplicity as that used in MP2. It must also, as operator, be a function of the wave function $\Psi^{(0)}$, but not of the orbitals used in the CASSCF that produced the wave function. Identically, the same wave function can in principle be represented by many different orbitals, as long as the CI coefficients are transformed together with the orbitals. Attempting to get a stable definition of orbitals, one may use, for example, natural orbitals or quasi-canonical orbitals. However, one finds in practice that during, for example, a geometry optimization, the natural occupation numbers of the excited state one is interested in undergo a near crossing. If one uses quasi-canonical orbitals, the orbital energies may cross. The unavoidable result is that the orbitals used in the CASSCF are not good variables for determining an $\hat{H}_0$. If the geometry varies, we must require that coming back to (almost) the same geometry will give (almost) the same computed correlation energy.

The solution we have used is to devise $\hat{H}_0$ as a function of $\Psi^{(0)}$ and $\hat{H}$, regardless of orbital representation. In any specific set of orthonormal orbitals, we may define the operator $\hat{F}$ as with elements

$$F_{pq} = \left\langle \Psi^{(0)} \left| \left[\hat{a}_p, \left[\hat{H}, \hat{a}_q^\dagger \right] \right]_+ \right| \Psi^{(0)} \right\rangle, \tag{12.8}$$

$$\hat{F} = \sum_{pq} F_{pq} \hat{a}_p^\dagger \hat{a}_q. \tag{12.9}$$

The matrix elements will transform correctly with orbitals in such a way that, if $\hat{H}$ was a one-electron operator, $\hat{F}$ would reproduce it (apart from some possible constant). The operator has eigenvalues with a property similar to that of Koopmans' theorem: the eigenvalues are approximate ionization energies and electron affinities (with reversed sign), taking an appropriate weighted average for active orbitals with fractional occupation. The resulting Fock matrix structure is shown in Figure 12.2.

This definition would be appropriate for a UHF formulation; for a spin-free operator, we average over the spin components M_S of the spin manifold of which $\Psi^{(0)}$ is one member; this is equivalent to average over α and β spin of the definition above. Evaluating the result using the second-quantized Hamiltonian gives a simple formula,

$$F_{pq} = h_{pq} + \sum_{rs} \left((pq|rs) - \frac{1}{2}(ps|rq) \right) D_{rs},$$

which is a well-known Fock-type matrix.

Internally in the CASPT2 program, we would like to use this operator as our $\hat{H}_0$; transformations among the inactive, active, and virtual orbitals (together with the appropriate transformation of CI coefficients to keep $\Psi^{(0)}$ unchanged) is used to bring the matrix **F** to be diagonal within these three subspaces.

However, like any one-electron operator must, it still couples any many-configuration $\Psi^{(0)}$ with the other wave functions in the CAS space, and in general, it also has nonzero elements that couple inactive, active, and virtual orbitals. The first deficiency, as a putative $\hat{H}_0$ definition, is managed by realizing that this coupling is already known to be identically zero for the true $\hat{H}$, as long as the CI part of the CASSCF was converged, which we always assume. This property is operationally conferred to the CASPT2 equation system since $\Psi^{(1)}$ is kept orthogonal to $\Psi^{(0)}$, and

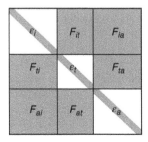

Figure 12.2   The CASPT2 Fock matrix, used to define $\hat{H}_0$. Gray areas: Nonzero matrix elements.

by computing the right-hand side as the projection of $\hat{H}\Psi^{(0)}$ onto the interacting space. This means that in any formal analysis, the $\hat{H}_0 - E_0$ will be used in the form

$$\hat{H}_0 \overset{\text{def}}{=} \hat{P}_{\text{CAS}}\hat{H}\hat{P}_{\text{CAS}} + \hat{P}_{\text{SD}}\hat{F}\hat{P}_{\text{SD}},$$

where $\hat{P}_{\text{CAS}}$ projects onto the CAS wave function space, while $\hat{P}_{\text{SD}}$ projects onto the interacting space generated by the excitation operators.

The coupling between the inactive, active, and virtual spaces through $\hat{H}_0$ can be similarly ignored, or it can be kept. In the first case, the submatrices for individual excitation cases, as indexed by K in the CASPT2 equations, Eq. 12.7, are uncoupled; we get a direct solution. Else, they are brought in by solving these equations iteratively, using the uncoupled direct solution as a preconditioner in a Preconditioned Conjugate Gradient (PCG) solver [10]. Early experiments showed that the coupling should be kept, in particular for state-averaged CASSCF, and since then the first approach has not been used.

12.1.6 IPEA, RASPT2, and MS-CASPT2

Already when CASPT2 was first implemented and tested, an almost-systematic defect was noted. For excitation energies relative to a closed-shell ground state, there appeared to be a few kcal/mol error proportional to the number of pairs that was broken in the excited state. This was attributed to the use of the same orbital energies in the active space, regardless of if an excitation transferred electron to, or from, an orbital. The major effect (the selection effect mentioned above) is taken care of by the explicit calculation of energy values in CASPT2. However, there is a remaining effect that has had to be treated by a more heuristic approach. Early experiments by Andersson [11] and other resulted in the correction models dubbed g_1, g_2, and g_3. They each gave systematic corrections of the right magnitude but finally lost the competition with Ghigo's IPEA shift [8], where a correction is added to the active orbital energies. The correction depends on a global parameter value, on the active density matrix elements, and whether the excitation case ("VJTU" etc.) is exciting active electrons into, or out from, the active space.

An optimal value of the global parameter (called the "IPEA shift") of 0.25 was found to be generally successful in reducing errors for atoms and diatomic molecules across the periodic table, and was adopted as a default choice in our programs. It has been found later that this choice is not universally successful, and that a larger value is indicated for some transition metal systems [12, 13]. On the other hand, later studies suggest that the demand for large IPEA shifts was exaggerated by an insufficiently large active space. The jury is still out on that question.

In the case of RASSCF root functions, a possible coupling between RAS1, RAS2, and RAS3 subspaces cannot be eliminated by diagonalizing the active/active part of the Fock matrix, since this would alter the $\Psi^{(0)}$ wave function. Such coupling would have to be treated iteratively in the PCG, and this would require a much higher computation load and complicated coding, including four-particle density matrices over the active orbitals. This goes contrary to the idea of RASPT2, which is to allow use of larger active spaces. The rational solution seems to be to ignore such coupling and regard them as a remaining inaccuracy resulting from using the RASSCF as an approximation to the full CASSCF.

As described so far, CASPT2 is a multiconfigurational but single-reference method. However, once the first-order wave functions have been obtained, each has its own zeroth-order Hamiltonian. These define individual reduced resolvents, that is, the linear mapping from $\Psi^{(0)}$ to $\Psi^{(1)}$ is different for each of several root states. As argued by Zaitsevskii and Malrieu [14], this defines a "multiresolvent," and by regarding the selected space of root functions as the internal space of a quasi-degenerate perturbation theory, the Multistate CASPT2 (MS-CASPT2) is obtained [15]. An effective Hamiltonian is constructed, where the diagonal elements are the single-state CASPT2 energies, while the nondiagonal states are computed as matrix elements of the first-order wave functions. This computation is time-demanding if there are many root states. Each nondiagonal element of the effective Hamiltonian requires the two interacting wave functions to be transformed to a common set of orbitals. Finally, the effective Hamiltonian is symmetrized and diagonalized.

This procedure has been found useful when different root functions are dissimilar but produced from a single state-averaged RASSCF. It can to some extent alleviate the problem that occurs when the energy curves of two states have an avoided crossing, if this occurs at a different geometry than the CASSCF energies, but recently a method called Extended Multistate CASPT2 (XMS-CASPT2) has emerged as probably a better alternative for such situations. This procedure uses for each state the same Fock operator, computed as an average for several states of a State-Averaged CASSCF. It is not yet clear if this is the preferred alternative in general (The common Fock matrix could then be a compromise that may be less efficient than the use of individual resolvents.), but it is definitely to be preferred in cases of weak avoided crossings or conical intersections.

12.2 REFERENCES

[1] Andersson K, Malmqvist PÅ, Roos BO. Second-order perturbation theory with a complete active space self-consistent field reference function. J Chem Phys 1992;96:1218–1226.

[2] Andersson K Multiconfigurational Perturbation Theory [PhD thesis]. Lund University Department of Theoretical Chemistry, Chemical Center, P.O.B. 124, S-221 00 Lund, Sweden; 1992.

[3] Angeli C, Pastore M, Cimiraglia R. New perspectives in multireference perturbation theory: the n-electron valence state approach. Theor Chim Acta 2007;117:753–754.

[4] Nakano H, Nakatani J, Hirao K. Second-order quasi-degenerate perturbation theory with quasi-complete active space self-consistent field reference functions. J Chem Phys 2001;114:1133–1141.

[5] Chandhuri RK, Freed KF, Hose G, Piecuch P, Kowalski K, Włoch M, Chattopadhyayd S, Mukheriee D, Rolik Z, Szabados A, Toth G, Surjan P. Comparison of low-order multireference many-body perturbation theories. J Chem Phys 2005;122:Art. 134110.

[6] Ghosh A, Gonzalez E, Tangen E, Roos BO. Mapping the d-d excited-state manifolds of transition metal-diiminato-imido complexes. A comparison of DFT and CASPT2 energetics. J Phys Chem A 2008;112:12792–12798.

[7] Schreiber M, Silva-Junior MR, Sauer SPA, Thiel W. Benchmarks of electronically excited states: CASPT2, CC2, CCSD, and CC3. J Chem Phys 2008;128:Art. 134110.

[8] Ghigo G, Roos BO, Malmqvist PÅ. A modified definition of the zeroth order Hamiltonian in multiconfigurational perturbation theory (CASPT2). Chem Phys Lett 2004;396:142–149.

[9] Dyall KG. The choice of a zeroth-order Hamiltonian for second-order perturbation theory with a complete active space self-consistent-field reference function. J Chem Phys 1995;102:4909–4918.

[10] Hestenes MR, Stiefel E. Methods of conjugate gradients for solving linear systems. J Res Nat Bur Stand 1952;49:409–436.

[11] Andersson K. Different forms of the zeroth-order Hamiltonian in second-order perturbation theory with a complete active space self-consistent field reference function. Theor Chim Acta 1995;91:31–46.

[12] Kepenekian M, Robert V, Guennic BL. What zeroth-order Hamiltonian for CASPT2 adiabatic energetics of Fe(II)N(6) architectures? J Chem Phys 2009;131:Art. 114702.

[13] Daku LML, Aquilante F, Robinson TW, Hauser A. Accurate spin-state energetics of transition metal complexes. 1. CCSD(T), CASPT2, and DFT study of [M(NCH)6]2+ (M = Fe, Co). J Chem Theory Comput 2012;8:4216–4231.

[14] Zaitsevskii A, Malrieu JP. Multi-partitioning quasidegenerate perturbation theory. Chem Phys Lett 1995;223:597–604.

[15] Finley J, Malmqvist PÅ, Roos BO, Serrano-Andrés L. The Multi-state CASPT2 method. Chem Phys Letters 1998;288:299–306.

13

CASPT2/CASSCF APPLICATIONS

In the following, we provide the reader with a series of example of how the CASPT2/CASSCF paradigm is used in a number of chemical problems. These examples might be of the trivial nature of molecular structure, to more sophisticated investigations on excitation spectra, photochemistry, dissociation processes, transition metal chemistry, and lanthanide and actinide chemistry. In some cases, examples are given in which so-called spin-orbit coupling effects cannot be ignored. Each example carefully defines the difficulty of the investigation and puts this into the perspective of selecting an active space for subsequent CASPT2/CASSCF investigations. In doing so, we follow the rules as put forward previously, but we also deal with some exceptions. In addition, we try to communicate some of the tricks that an experienced CASSCF user would utilize in times of troubles.

In common, all these examples have the delicate task of selecting the active space. As previously mentioned, some semigeneral rules can be proposed; however, they do not always apply. The general rule, if there is one, is that the investigator has to understand the chemistry to be modeled at the molecular orbital level. In doing so, an appropriate active space can be selected. It is worth noting that this can in some cases lead to the same molecular system is studied with two different active spaces depending on the purpose of the study. For example, to study the lowest excited states of an unsaturated system, only the π orbitals have to be included in the active space, whereas if the dissociation of the same system is to be studied this calls for the inclusion of the σ orbitals too. In addition, the selection of the active space is not simply a

Multiconfigurational Quantum Chemistry, First Edition.
Björn O. Roos, Roland Lindh, Per Åke Malmqvist, Valera Veryazov, and Per-Olof Widmark.
© 2016 John Wiley & Sons, Inc. Published 2016 by John Wiley & Sons, Inc.

matter of determining the *number* of active electrons and orbitals. It is also a matter of them having the correct character. The CASSCF model is a complicated mathematical model which, for a given molecular structure, spin, overall symmetry, and number of active electrons and orbitals, can have multiple valid stationary solutions. To simply use canonical SCF orbitals as a start in the CASSCF procedure and magically expect these to generate the correct active space is just a demonstration of the ignorance of the understanding of the complexity of the CASSCF model. Hence, the prudent user of the CASSCF method will *always* check that the orbitals of the converged CASSCF wave functions are consistent with the intentions. If not she or he will have to go back to the drawing board and reconsider the reason behind the design of the active space. In this respect, a molecular orbital graphical viewer interfaced to the particular implementation of the CASSCF procedure in use is instrumental toward the ability to secure the consistency of the performed compilations.

Before we start with the individual examples below, we spend some time on methods and tools to generate sets of orbitals from which we can select the starting orbitals. These tools are as valuable and important to the successful use of the CASSCF method as completing the intellectual challenge to select the active space for the particular molecular system and chemical process under study.

13.1 ORBITAL REPRESENTATIONS

The selection of active orbitals results in the end with the actual generation of a set of starting orbitals for the CASSCF procedure. This problem is more of a technical nature than anything else. However, without this technical ability, much of the usefulness of the CASSCF method is lost. We noted previously that the CASSCF wave function equations can have multiple solutions for a given selection of active orbitals and electrons. In this respect, we need the initial selection of the starting orbitals to be a close as possible to what we believe to be the correct model space if we want to assure convergence to this model space.

It has been a tradition to select the initial starting vectors for CASSCF calculations from SCF canonical orbitals. This has been rather a successful approach, but it also has its limitations for two reasons. First, the canonical SCF orbitals tend to be delocalized (see Figures 13.1 and 13.2) and canonical virtual SCF orbitals can occur in the wrong order as compared to the "standard" HOMO-LUMO order (compare the virtual orbitals of Figure 13.1 with the orbitals in Figure 13.3). The SCF representation of the virtual orbitals is in conflict with the natural way in which we conceive and construct the active space. The most natural way to envision the active orbital space is in the mode of atomic and lone-pair orbitals, or with localized bonding and antibonding orbitals and lone-pair orbitals. Second, the virtual orbitals in the SCF procedure are to some extent just a complement to the occupied orbital space and in this sense lack any significant characters that can be associated with atomic or localized molecular orbitals required in the design of an active space. This problem is exaggerated by (i) the used of extensive basis sets beyond the double zeta quality (see Figure 13.3) and (ii) the fact that we today can apply the

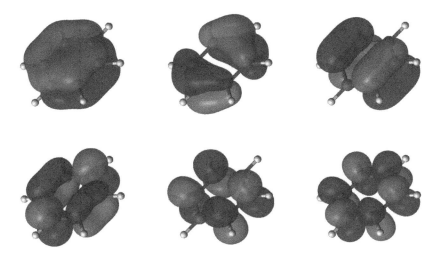

Figure 13.1 The occupied (bottom row) and virtual (top row) π canonical SCF molecular orbitals of benzene, from a calculation with no explicit symmetry constraints. In a calculation with a minimal basis (ANO-RCC). This set of orbitals constitute the HOMO-4, HOMO-1, HOMO, LUMO, LUMO+1, and LUMO+2 SCF orbitals of the molecular system. Note how these orbitals are delocalized over the whole conjugated system.

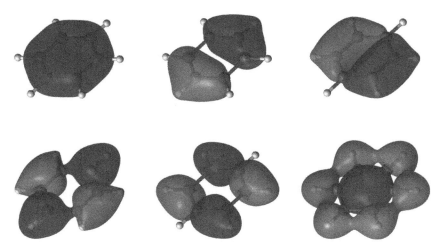

Figure 13.2 The energetically six lowest occupied σ canonical SCF orbitals of benzene, from a calculation with no explicit symmetry constraints and a minimal basis (ANO-RCC). Here we note that the canonical SCF molecular orbitals are not only delocalized but also that they are a mixture of CC and CH σ bonds. This makes these orbitals rather useless as starting orbitals for a CASSCF in which, for example, one would like to study the hydrogen abstraction reaction.

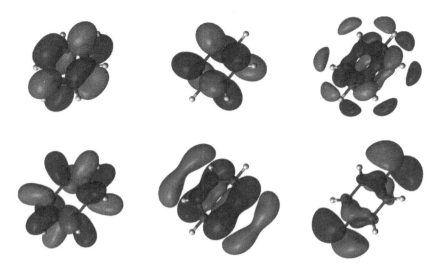

Figure 13.3 The energetically six lowest virtual π^* canonical SCF orbitals of benzene, from a calculation with no explicit symmetry constraints in conjunction with a basis set of triple ζ quality (ANO-RCC). The first observation is that these orbitals are not found among the three lowest virtual orbitals. The LUMO and the following virtual orbitals can best be described as garbage orbitals to fulfill the orthogonality conditions of the canonical virtual orbitals, for all what concerns us they completely lack any character that will help in the selection of starting orbitals (compare with the upper row in Figure 13.2). The second observation is that these π^* orbitals contain a heavy mixture of other types of antibonding orbitals, again rendering these useless for the purpose of generating CASSCF starting orbitals.

CASPT2/CASSCF method to much larger systems as compared to in the past. Both these facts will generate virtual orbitals that are rather useless for starting orbitals in the CASSCF procedure. Hence, tools are needed to generate good starting orbitals that goes beyond a simple use of SCF orbitals. But, before resorting to that problem, let us address the problem with the virtual orbitals in association with large basis set. This problem is twofold. First, finding the appropriate starting orbitals gets cumbersome as the virtual orbital space grows with the size of the basis set. Second, with a growing orbital basis, the virtual orbitals look more and more exotic and do not have any meaningful characteristics. This problem is solved with the following recommendation.

- Generate starting orbitals in a *minimal* basis set, that is, with a basis limited in size to core and valence orbitals. In this basis set, the atomic orbitals and molecular orbitals are produced with characteristics with which one can easily associate chemical qualities as lone-pair, radical, bonding, nonbonding, and antibonding orbitals. Especially, the minimal basis will generate virtual orbitals that are rather easy to assign and identify. These properties are very suitable in the design of the orbitals of the active space.

- Produce converged CASSCF orbitals in the minimal basis set. Inspect that the orbitals are consistent with the intentions. Pay special attention to the orbitals that have natural occupation numbers close to 2.00 or 0.00.
- Expand the CASSCF orbitals in the minimal basis set on a larger basis set. Expand a minimal on double, a double on triple ζ quality basis set, etc. Again, inspect the converged CASSCF orbitals.

The experience we have from this procedure over the last few years has been more than satisfactory. The only problem with this approach is the possibly emerging problems associated with Rydberg states. As described below, calculations on Rydberg states require special care and, in particular, specially designed Rydberg basis sets are required. However, alternatively, as the atom-centered basis set is expanded with more and more diffuse functions its ability to describe Rydberg orbitals increase – although far from perfect. Sometimes, for example, when there is an accidental degeneracy between an excited valence and Rydberg state, the expansion to a larger basis set will lead to an instability of the active space if not both orbitals to describe the excited valence orbitals and the Rydberg orbitals are included in the active space.

Let us now turn our attention to the generation of the initial set of starting CASSCF orbitals in a modestly or minimally sized basis set. As have been discussed in previous chapters, the active orbitals are selected from the perspective of solving a particular chemical process as acting on a molecular system. These considerations sometimes are best conducted in terms of atomic orbitals, other times molecular orbitals are more appropriate. Fortunately, various schemes for orbital localization have been developed over the years to reduce the scaling in *ab initio* implementations, in particular for schemes that include electron correlation. These localization schemes are also important to the design of atomic and localized molecular orbitals to be used as starting vectors in the CASSCF procedure. In the following, we use the localization schemes by Pipek–Mezey [1], the projected atomic orbital (PAO) procedures [2, 3] and the so-called Cholesky localized orbitals by Aquilante et al. [4] Below it is demonstrated and given examples on how the localization schemes can be used to generate useful starting orbitals. This will be done for the trans-butadiene molecular system (see Figure 13.4). The localization scheme by Edmiston–Ruedenberg [5] will

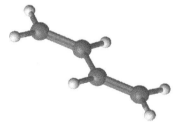

Figure 13.4 The trans-butadiene molecule. The carbon atoms of this conjugated system are sp^2 hybridized. The remaining carbon unhybridized 2p orbitals form a set of orbitals perpendicular to the plane of the molecular system. This set of orbitals constitute the orbital subspace of the conjugated orbitals.

not be used since this localization scheme minimizes the self-interaction energy and uses the two-electron integrals in the localization scheme. The scheme by Boys [6] is not used since it does not preserve the separation of the σ and π space. In that respect, it is not useful to the design of the orbitals of the active space for the CASSCF model. In the trans-butadiene case, the selection of the active space will be viewed for three different processes:

- low-lying valence electronic excitations,
- a terminal-hydrogen abstraction, and
- the fragmentation process into two equivalent vinyl moieties.

13.1.1 Starting Orbitals: Atomic Orbitals

Out of the three chemical processes, the first one deals with the low-lying valence excitations. The low-lying excitations of the trans-butadiene molecule are of π to π^* character, that is, excitations within the orbital subspace spanned the conjugated p-orbitals. The associated CASSCF active space to correctly describe these excitations can be done in either atomic or molecular orbitals. The valence excitations of trans-butadiene will here be used as a trivial example to demonstrate the use of atomic orbitals to design an active space. The use of atomic orbitals in this case could be argued to be redundant in comparison with a straightforward approach using the SCF molecular orbitals. This is certainly the case in which the active space is "complete," that is, it includes all the relevant orbitals like in a full-valence or full-π CI. However, from time to time, this will not be possible and a truncation of the active space will have to be introduced. The truncation could be executed to remove orbitals with occupation numbers close to 2.00 or 0.00. However, in some cases, localization is more suitable, for example, the twisting energy profile of a long poly-alkane chain can be investigated using a reduced active space including only the atomic conjugated orbitals around the point of the twist.

For trans-butadiene, the conjugated orbital subspace has its origin from the atomic carbon 2p orbitals perpendicular to the molecular plane (see Figure 13.5), and an appropriate set of starting orbitals for the CASSCF active space could be generated by including these atomic orbitals. The procedure that was used here was as follows:

- An initial set of approximative starting molecular orbitals were generated with SCF adding four electrons to make sure that all the π orbitals were occupied.
- These orbitals were orthonormalized by the PAO or Pipek–Mezey localization method.

The resulting starting orbitals contain molecular orbitals describing the σ bonding orbitals followed by a set of orthonormal atomic orbitals associated with the conjugated system. The definition of these atomic orbitals are trivial if one uses ANO type

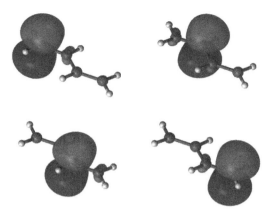

Figure 13.5 The orthonormal atomic 2p carbon orbitals of trans-butadiene, which are the building blocks of the conjugated π systems. These orbitals have been generated from an initial set of atomic orbitals, which through the PAO procedure have been made orthonormal to each other.

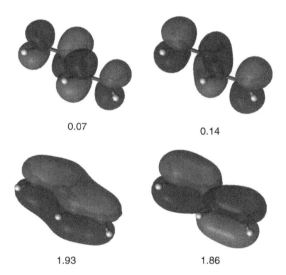

0.07   0.14

1.93   1.86

Figure 13.6 The converged natural orbitals of the CASSCF active space designed from the orbitals of the conjugated system, that is, the π (bottom row) and π^* (top row) orbitals. The values below each set of orbitals represent the partial natural occupation number.

of basis sets. The PAO or Pipek–Mezey procedure is found to work equally well and produce starting orbitals, which under visual inspection, are identical to the eye. The converged CASSCF orbitals will be identical regardless of the localization scheme used in the generation of the initial set of starting orbitals (see Figure 13.6).

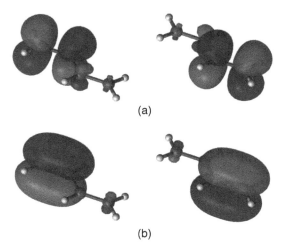

Figure 13.7   The localized Pipek–Mezey molecular orbitals of the conjugated system, the π bonding (b) and π antibonding (a) orbitals.

(a)   (b)

Figure 13.8   The localized Pipek–Mezey σ (a) and σ^* (b) orbitals of the terminal CH bond.

13.1.2   Starting Orbitals: Molecular Orbitals

First, use the localization scheme to generate molecular orbitals for the active space of a study of the lowest valence-excited state, π to π^* excitations. Here the starting orbitals are generated by applying the Pipek–Mezey localization scheme to the SCF orbitals (see Figure 13.7), first on the occupied orbital subspace and then on the virtual orbital subspace, respectively. The SCF procedure separates the original atomic orbital manifold into occupied and virtual subspaces. The subsequent localization scheme will in an elegant way generate localized bonding and antibonding orbitals. The use of these localized orbitals will again render a converged set of CASSCF orbitals identical to those of Figure 13.6.

In the next case, an active space will be selected to facilitate the study of the hydrogen abstraction of one of the terminal hydrogens. As the CH bond is extended, the σ and σ^* orbital of that particular CH bond (see Figure 13.8) will become degenerate and should for this reason be included in the active space if one would like to have a qualitatively correct wave function description of the bond breakage. In this particular case, the π space orbitals could be left outside the active space. However, it should

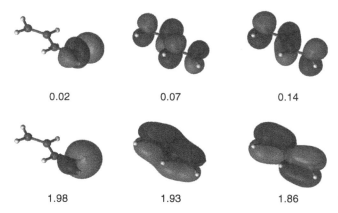

0.02   0.07   0.14

1.98   1.93   1.86

Figure 13.9   The converged natural orbitals of the CASSCF active space to study the hydrogen abstraction process of a terminal hydrogen in the trans-butadiene molecule. The lower row represents the bonding orbitals and above each of these orbitals, the corresponding correlating orbital is depicted. The number below each orbital represents the natural occupation number. Note that for each pair of correlating orbitals the occupation numbers add up to 2.00.

be included in the active space since the degeneracy of the π, here and in general, often results in a strong multiconfigurational wave function, that is, an SCF description of this subspace is far from a quantitatively correct description. Here a short discussion of the consistency of this active space might be in place. The prudent user of the CASSCF method will at once protest against just including *one* CH bond in the active space. This would render the CASPT2/CASSCF optimized structure not to be symmetric. Hence, arguments could be made why at least the symmetric partner of this particular CH bond should be included in the active space. Thus, a more strict CASSCF practitioner would argue that *all* CH bonds should be included, or at least the orbitals of the symmetry-related CH bond (this would ensure that symmetry of the molecule is preserved before the dissociation). This will, although, in the case of larger systems be prohibitive and no CASSCF calculations would be possible in practice. It is noted, however, that philosophical problems such as these normally are only observed in systems with symmetry and that in most cases a more pragmatic and restrictive view can be adopted toward enlargements of the active space for just pure symmetry reasons. Some examples below in this chapter explicitly address this question. Nevertheless, the converged natural orbitals of the CASSCF active space with six electrons in six orbitals are displayed in Figure 13.9.

Finally, we turn our attention to the case in which the trans-butadiene molecule is fragmented into two vinyl radicals. Here it is noted that, in addition to a breakage of the CC σ bond, that a π bond is broken too. In that respect both the π subspace (see Figure 13.7) and the σ bonding and antibonding orbital of the center CC bond has to be included in the active space(see Figure 13.10). The natural orbitals of the corresponding active space, six electrons in six orbitals (6-in-6), are presented in Figure 13.11. Note that this active space with respect to the one for the hydrogen

(a)   (b)

Figure 13.10   The σ (a) and σ^* (b) orbitals of the central CC bond of trans-butadiene.

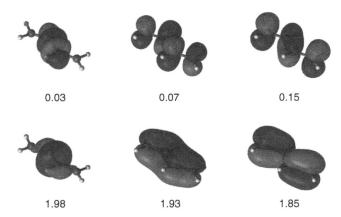

0.03   0.07   0.15

1.98   1.93   1.85

Figure 13.11   The converged natural orbitals of the CASSCF active space for the study of the fragmentation of the trans-butadiene molecule into two vinyl radicals. The number below each orbital represents the natural occupation number.

abstraction has *the exact same number of electrons and orbitals*. Still with different sets of starting orbitals *two different stable solutions* to the CASSCF equation were found. This demonstrates in a clear way that a deep understanding of the chemical problem on a molecular orbital level of theory and the selection of the start orbitals are instrumental to a successful use of the CASSCF method. This part is wrapped up by using the converged starting orbital of this case, as generated in a minimal basis, and expand it up to a triple ζ basis set (see the top of this section). The final results are found in Figure 13.12. Here it is noted that the character of the orbitals are almost identical and that the change of basis set only has affected the occupation of the π orbitals close to the Fermi gap.

Finally, it is noted that the procedures to generate the starting orbitals for CASSCF active spaces are based on the experience and developments of the authors of this book. In this respect, it is expected that other groups have generated other modes of generating the starting orbitals. Which method is the best could be argued. However, it is the firm message of this section to communicate to the reader that some quality time has to be spent on the generation of the starting orbitals in the CASSCF approach to produce meaningful results. The authors of this book have come to the conclusion that this is time well spent.

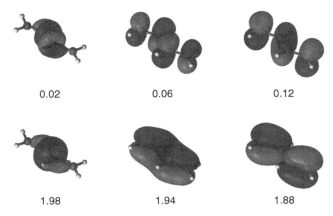

0.02   0.06   0.12

1.98   1.94   1.88

Figure 13.12 The converged natural orbitals of the CASSCF active space for the study of the fragmentation of the trans-butadiene molecule into two vinyl radicals in a triple ζ quality basis set. The number below each orbital represents the natural occupation number.

With this technical problem now discussed, the attention is not turned to some examples from some actual CASPT2/CASSCF studies. Once again, the set of examples that follows has been selected to cover CASPT2/CASSCF studies from as many parts of the periodic table and as many different chemical processes as possible.

13.2 SPECIFIC APPLICATIONS

In the following, some explicit examples are given. Most cases are taken from the literature; however, the review to follow is in no way intended to replace the reading of the original papers. Rather, the major task at hand in this chapter is the selection of the active space in all the individual cases. This review does not primarily focus on the results of the studies from which these examples have been gathered. All the cases are dealt in a uniform way, hence all cases contain a brief introduction to the chemistry, a discussion on the selection of the active space, and a summary with some of the results. All examples are supported with numerous figures.

13.2.1 Ground State Reactions

The Bergman Reaction

Introduction The so-called enediyne class of anticancer drugs have been of interest for the last 15 years, as a potent new agent in the human quest to tackle the growing problems of cancer in today's society. The calicheamicin molecule (see Figure 13.13), discovered in 1987, displayed high potency against cancer cells. The molecular system contains tethers (left part of the molecule) that will position the molecule in the minor groove of DNA and drive in the warhead, an enediyne moiety (see the right

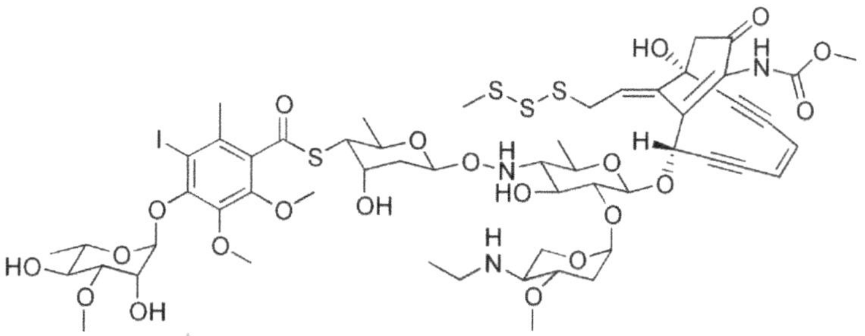

Figure 13.13 The calicheamicin molecule discovered in 1987 in bacteria living in chalky rocks.

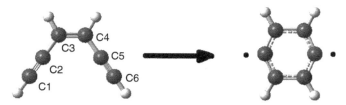

Figure 13.14 The Bergman cyclization reaction, the transformation of (Z)-hex-3-1,5-diyne into the *para*-benzyne molecule.

side of the molecule), in between the two strands of the double helix. This will be followed by a so-called trigger reaction that will activate the warhead to undergo a so-called Bergman cyclization [7–10] (see Figure 13.14) and produce a biradical intermediate. This will subsequently attack both sides of the double strand, abstracting two hydrogen atoms. Ultimately, this process leads to DNA cleavage and the death of the host cell. This activity is for some unclear reason more active in cancer cells as compared to normal cells. To understand the process and modify the enediyne class of anticancer drugs to improve the efficiency the first type of *ab initio* studies of these systems were limited to studies of the Bergman cyclization reaction itself [11, 12]. Understanding the details of this process would put the scientist in control of parameters that could govern the reaction and activation energies of this important reaction.

In addition, the Bergman reaction is a challenge to DFT theory. During this reaction, the wave function will change character from closed shell to a singlet biradical character. The former can be modeled with closed-shell DFT while the latter would require open-shell DFT. This will enter two complications for the DFT model: (i) When will the closed-shell reference density break into an open-shell reference density – can this be modeled in a consistent way? (ii) The open-shell reference density will be a mixture of a singlet and a triplet – will the spin-contamination be of any significance? For the CASSCF/CASPT2 approach, this is no problem. First, the wave function will have no problem to model the gradual change of the wave function from closed shell to open shell. Second, the CASSCF wave function is a spin eigenfunction.

The Active Space   The Bergman cyclization reaction can in some sense be viewed as an internal Diels–Adler reaction, a set of π orbitals rearranged into σ bonds upon a geometry distortion. In the case of the Bergman reaction, the trivial active space would involve for the reactants and the products the four orbitals of the in-plane π system, and the $C_1 - C_6$ σ and σ^* and the two radical orbitals of carbons C_2 and C_5, respectively. However, the formation of a conjugated system in the enediyne to an aromatic system in the *para*-benzyne structure needs to be considered. Hence the selection of the extended active space seems simple. We have an out-of-plane π system with near-degeneracy effects (see the three leftmost set of orbitals of the top and center row of Figure 13.15, and the top row of Figure 13.16) and we are

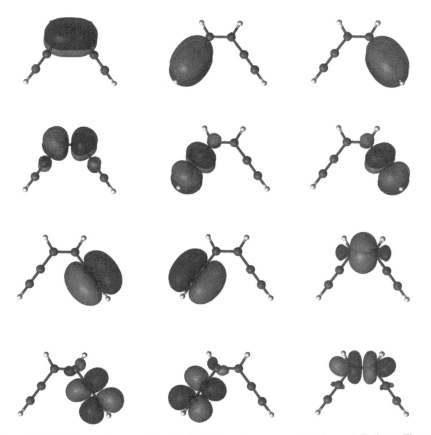

Figure 13.15   The starting atomic orbitals of the active space of (Z)-3-ene-1,5-diyne. The top two rows display the out-of-plane π orbitals and π^* orbitals, respectively; the first two orbital sets on the third and fourth row display the in-plane π and π^* orbitals. The last orbital set, on the same rows, displays the σ and σ^* orbitals of the C_3–C_4 bond.

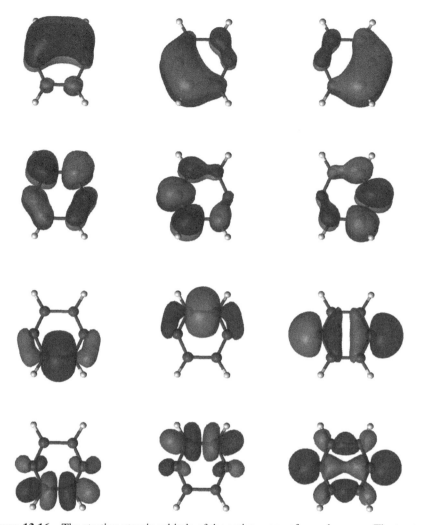

Figure 13.16 The starting atomic orbitals of the active space of *para*-benzyne. The top two rows display the out-of-plane π and π^* orbitals, respectively; the first two orbital sets on rows three and four display the σ and σ^* orbitals. The last orbital set, on the same rows, displays the two radical orbitals.

simultaneously forming two radical orbitals (see bottom row of Figure 13.16) and a σ bond (see orbital sets 1 and 3 of the center row of Figure 13.16) from the in-plane π orbitals of the triple bonds ($C_1 - C_2$ and $C_5 - C_6$) (see the two leftmost orbital sets of the top and center row of Figure 13.15). This would naturally call for that we had the six out-of-plane p orbitals of the π system and the four orbitals of the in-plane π and π^* orbitals of the triple bonds of the (Z)-hex-3-ene-1,5-diyne in the active space. The former would carry six and the latter four electrons. However, the degeneracy of the newly formed bond in the *para*-benzyne, $C_1 - C_6$, with the $C_3 - C_4$ bond of the enediyne requires that bonding and antibonding orbital of this bond is included too (see center row of Figure 13.16). If not the CASSCF or CASPT2 optimized structure will display a symmetry-broken solution for the *p*-benzyne structure. Hence, these orbitals should be included in the active space too. One could argue that the remaining carbon–carbon σ bonds should be included too in order to completely eliminate any bias. Indeed, a CASSCF optimized structure will give too long $C_1 - C_6$ and $C_3 - C_4$ bonds while the other will be too short. However, at the CASPT2 level of theory, this bias will to a large extent be eliminated, and with the present limitations of the CASSCF implementations we have to make a compromise. The resulting CASSCF active natural orbitals and corresponding natural orbital occupation numbers of the enediyne and *para*-benzyne structure are displayed in Figures 13.17 and 13.18. In particular, we note how the starting atomic orbitals now have transformed into molecular orbitals. This is especially the case for the π orbitals while the starting σ orbitals are very similar, apart from the formation of symmetric and antisymmetric recombination for the *para*-benzyne, to the final natural orbitals. We also note that the occupation number for the bonding orbitals are in the range of 1.89–1.95, which clearly demonstrates the closed-shell character of the electron structure of the enediyne. Note that total occupation number for each set of correlating orbitals almost perfectly add up to two electrons, just as we would expect! For the *para*-benzyne we note a similar trend, that is, the π starting atomic orbitals are transformed into molecular orbitals. In the π space, we again find a more or less perfect pairing of correlating orbital pairs (see top two rows of Figure 13.18). Finally, for the radical orbitals (see bottom row of Figure 13.18) we note the biradical character expressed by the occupation number of these two orbitals of 1.35 and 0.66, respectively.

Summary The Bergman reaction is a trivial example of a case in which symmetry considerations are important to avoid artifacts in the predicted optimized CASSCF geometry. The final active space of this system has to include 12 electrons in 12 orbitals; for further details in the literature we suggest the reader to consult [11, 12]. This reaction is also a case which in a beautiful way exhibits the power of the multiconfigurational method, which allows the wave function in a seamless manner to transform from a closed shell to a biradical electron structure as the structure of the molecular system is perturbed.

13.2.2 Excited States – Vertical Excitation Energies

This section provides two example of studies, on water and *para*-nitroaniline, in which the vertical excitation energies are studied with the CASSCF method. Both cases are to demonstrate how the active space has to be selected such that the implicitly defined CI expansion will include the desired excited states. In both cases, it will be seen how the active space is designed to model both valence $\pi - \pi^*$ and $n - \pi^*$ excitations and states of Rydberg character.

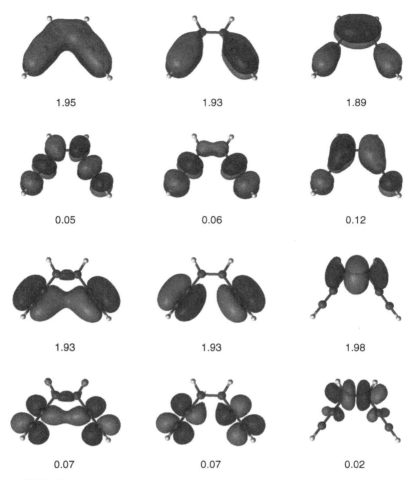

Figure 13.17 The natural orbitals of the active space and the natural orbital occupation numbers for (*Z*)-hex-3-ene-1,5-diyne treated with 12 electrons in 12 orbitals CASSCF. The top two rows display the out-of-plane π and π^* orbitals and occupation numbers, respectively, the first two sets of orbitals on rows three and four display similarly the in-plane π space. Finally, the last orbital set, on the same rows, displays the σ and σ^* orbitals of the C_3–C_4 bond.

Excited States of the Water Molecule: Analysis of the Valence and Rydberg Character

Introduction The study of excitation energies is problematic when excited states of both valence and Rydberg character is present in the same energy range. Due to the lack of dynamic correlation in the CASSCF model, artificial accidental degeneracies between these two types of excited states are created, which the CASSCF model cannot resolve. The only way to resolve these cases is to (i) include specifically designed

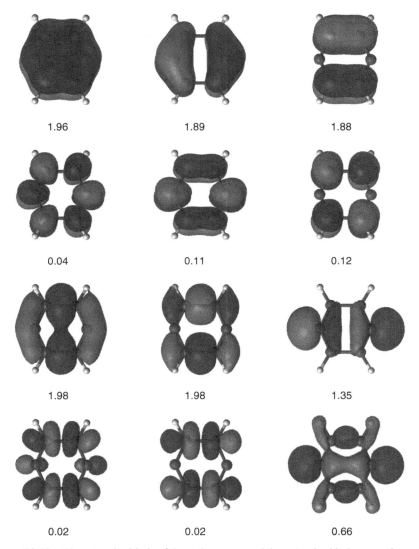

1.96   1.89   1.88

0.04   0.11   0.12

1.98   1.98   1.35

0.02   0.02   0.66

Figure 13.18   The natural orbitals of the active space and the natural orbital occupation numbers for *para*-benzyne treated with 12 electrons in 12 orbitals CASSCF. The top row displays the out-of-plane π and π^* orbitals and occupation numbers, respectively. The first two orbital sets on rows three and four display the σ manifold. Finally, the last orbital set, on the same rows, displays the radical orbitals.

molecular Rydberg basis sets, (ii) include in the active space both the important valence orbitals (occupied orbitals and the corresponding correlating orbitals) and the intruding Rydberg orbitals, and (iii) to resolve the artificial degeneracy with the MS-CASPT2 method. An excellent example of the use of the MS-CASPT2 method to resolve such problems is presented in the study of the V-state of ethene [13]. However, here we are going to demonstrate the case from a study of the low-lying excited states of water [14].

The Active Space   First let us establish a much overlooked fact – Rydberg orbitals are molecular in character and should not be generated from adding diffuse functions to an excising atomic basis set. Strictly speaking, one can do this of course, however, long before such a scheme converges the one-particle basis set will become linear dependent causing the basis set to be unnecessarily large and the optimization problem to be ill conditioned. Moreover, the standard basis sets are usually designed such that the diffuse parts of the basis set should correlate or polarize the atomic basis set, thus being inappropriate for the description of Rydberg orbitals. Hence, in this study, we will use an explicit molecular Rydberg basis set [14]. In this study, the molecular Rydberg basis was chosen to be situated on the oxygen atom of the water molecule, which is an exception, since molecular Rydberg basis sets normally are centered on the center-of-charge of the associated cationic system. In this particular case, the difference is insignificant.

   While the previously described procedure to generate CASSCF starting orbitals, using localized occupied and virtual SCF orbitals, is pretty straightforward for valence-type active orbitals, it is not very helpful in generating the candidate molecular Rydberg orbitals to be placed in the active space. Here something more elaborate is needed. What will follow is an extension to this scheme that has been developed at our lab during the past years. However, before we explore this, let us consider what we are about to do. We want to explore the combined valence and Rydberg electronic excited states of the water molecule. In this respect, we like to consider an active space composed of the water valence orbitals (see Figure 13.4) and add to this a set of Rydberg orbitals (the 3s, 3p, and 3d series of Rydberg orbitals and possibly beyond). When doing this and including both valence correlating orbitals and diffuse Rydberg molecular orbitals, it is important to understand the difference of these and their purpose for being in the active space. While the correlating orbitals are mirrored around the Fermi gap of the molecular system –correlating orbitals of valence orbitals are placed just above the Fermi gap and then for lower semicore and core orbitals the corresponding correlating orbitals are found at higher and higher orbital energies – the Rydberg orbitals are found right around or below the Fermi gap. The primary reason for including the correlating orbitals in the active space is to resolve the near-degeneracy effect and get a wave function that dissociates correctly. A second reason for including orbitals into the active space is that we like the CI expansion, implicitly defined in the CASSCF from the active space, to be able to describe excited states that have the character of electron excitations to these orbitals. Some of the correlating orbitals can serve this purpose too, while the inclusion of the Rydberg orbitals is only to facilitate the description by the CI expansion of Rydberg states. Depending on the total occupation and nature of the valence subspace the onset of Rydberg excitations will kick in early or late. For example, for π systems, which normally are half filled (one electron is formally assigned to each atomic p orbital that constitutes the π system), a significant number of excitations from bonding to antibonding valence orbitals will be of lower energy than the lowest excitation to a Rydberg orbital. However, for water, the situation is quite different. As discussed before, the valence correlating orbitals are just two ($4a_1$ and $2b_2$ – the symmetric and antisymmetric combination of the σ^* OH bonds) and high in orbital energy (σ^* orbitals have significantly higher orbital energies as compared to the π^* orbitals). Having a brute rule that single excited states occur roughly in the order $n - \pi^*$, $\pi - \pi^*$, $n - \sigma^*$, and n-Rydberg, it should be of no surprise that in a system as water, with no π system, that (i) the lowest excited state is of rather high energy (7.50 eV) and (ii) that the Rydberg states show their presence early on (in water the three lowest states are a mixture of valence and Rydberg character). The early onset of the Rydberg states in the valence-excited subspace of states forms a further complication – Rydberg and valence-excited states are differently sensitive to the lack of dynamic electron correlation in the CASSCF method. Typically, valence states are more stabilized by the subsequent inclusion of the dynamic electron correlation in the CASPT2 model as compared to the Rydberg states. This problem could cause artificial degeneracies between Rydberg and valence states having the CASSCF approach mixing electronic configurations that should not be mixed. Trivial perturbation theory approaches to resolve this with the standard state-specific CASPT2 method will not be able to compensate for the fact that the reference wave function is marred by an erroneous mixture of electronic configurations. This problem can only be treated with the multistate CASPT2 method. Hence, in a study like this, it is evident that qualitative and quantitative results can only be achieved by the combination of the SA-CASSCF and MS-CASPT2 being applied to a number of low-lying states of valence, Rydberg, or a mixture of valence and Rydberg character.

Let us move on with the design of the first set of starting orbitals for the CASSCF. We will, as before, start with a minimal valence basis set and now also a molecular Rydberg basis set (an uncontracted 2s2p2d Gaussian basis set). An initial active orbital set for the valence orbitals is generated as described before – from localized SCF orbitals of the neutral molecular system. The inclusion of the Rydberg basis requires some considerations, however. A simplistic inclusion of the Rydberg basis in the preceding step could work, although it is our experience that such orbitals will be rather noncharacteristic and too diffuse. To find a more well-defined set of orbitals to select from we will do as follows. The Rydberg basis set is added to the basis set and used in a CASSCF calculation on the cation of the molecular system. In this calculation, we use the valence active space and remove the electron that corresponds to the energetically lowest ionization. The electron affinity of this system will polish the virtual orbitals to something meaningful. The orbitals will be selected by the means of some visualization tool. Since the Rydberg orbitals are rather sparse in character and extent in space, it is important to reduce the isosurface value and to extend the box in which value of the orbitals are evaluated. The resulting orbitals of such a calculation, a full valence CASSCF calculation of the ground state water cation with one electron removed from the $1b_1$ orbital (the lone-pair orbital perpendicular to the H-O-H plane), are depicted in Figure 13.19. In particular, we note the significant difference in the spatial extent between the two correlating orbitals, $4a_1$ and $2b_2$, and the Rydberg set of molecular orbitals. Considering that standard atomic basis set,

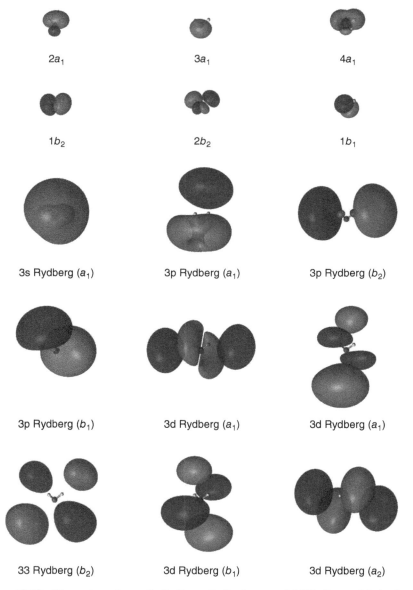

Figure 13.19 The water valence, 3s Rydberg, 3p Rydberg, and 3d Rydberg orbitals, from a full valence CASSCF calculation of H_2O^+ cation using a minimal valence and 2s2p2d molecular Rydberg basis set. The isosurface of the valence and Rydberg orbitals are 0.040 and 0.017 au, respectively.

augmented with diffuse functions, falls short of using the very diffuse functions of the molecular Rydberg basis set, it is questionable if these standard basis sets are sufficiently accurate to support investigations on Rydberg states. At this point, we can proceed with the calculation by extending the valence basis set up to the quality that we desire. Typically, for the computation of electronic excited states, we need a triple ζ quality or better. However, the size of the molecular primitive Rydberg basis set could be reduced by forming a contracted basis set, in accordance with the results from the calculation on the cationic system.

The subsequent SA-CASSCF calculations, which will follow after we have generated an appropriate set of orbitals, can be tailored in different ways. A trivial approach for water would be to make a calculation in which we add the 3s, 3p, and 3d sets of Rydberg orbitals to the valence orbital set in the active space. This would constitute 8 electrons in 15 orbitals. This is an unnecessarily too large CI (for water this amount to about half a million CSFs). Is there a way in which we can reduce this? Yes, indeed there is. Considering that the Rydberg orbitals are included in the active space not to introduce correlating orbitals to solve near-degeneracy effects but rather to assure that the CI expansion will include excitations to these orbitals we are free to only include individual Rydberg orbitals in accessing the excitation energy to this particular state. With this in mind, the calculation of the actual excitation energies can be performed in a series of calculations using a smaller active space. Here we will give an example in which we include the 3s and 3p Rydberg in the orbitals and compute the six lowest excited states of water. These states are the ground states followed by excisions to correlating orbitals, $4a_1$ and $2b_2$, and the 3s and 3p Rydberg orbitals. The calculation was performed as follows. The orbitals from the minimal valence basis set CASSCF calculation on the cation were used to select active orbitals for a 6 root SA-CASSCF with 8 electrons in 10 orbitals (the six valence orbitals followed by the four Rydberg orbitals). The converged SA-CASSCF orbitals were then used in a procure in which the valence basis set was expanded from a minimal basis to a triple ζ plus polarization quality. The resulting converged SA-CASSCF orbitals are displayed in Figure 13.20. First we note that the Rydberg orbitals are somewhat more contracted as compared with the orbitals generated initially in the cationic calculation. This is due to that the Rydberg orbitals are now occupied and then their shape is directly optimized as they now contribute to the optimized energy. Furthermore, the valence orbitals are slightly different since the valence basis set is now much more flexible. We also note that the SA-CASSCF natural orbital occupation numbers are consistent with the assignment of the lowest five excited states: $1b_1 \rightarrow 3s/4a_1$, $1b_1 \rightarrow 3p_y/2b_2$, $3a_1 \rightarrow 3s/4a_1$, $1b_1 \rightarrow 3p_z$, and $1b_1 \rightarrow 3p_x$ [14]. That is, there are two roots corresponding to excitations to the 3s Rydberg orbital, while there is just one state each corresponding to an excitation to a 3p Rydberg orbital. Moreover, the occupation numbers of the correlating orbitals are considerably lower than those for the Rydberg orbitals, indicating that these orbitals are not of significant importance to describe orbitals to which we are exciting the electrons but rather are just used for correlating purposes. The excitation energies are computed as follows. The ground state energy is established from a full valence single-root CASSCF followed by a state-specific CASPT2. The energies of the excited states are established in an SA-CASSCF over

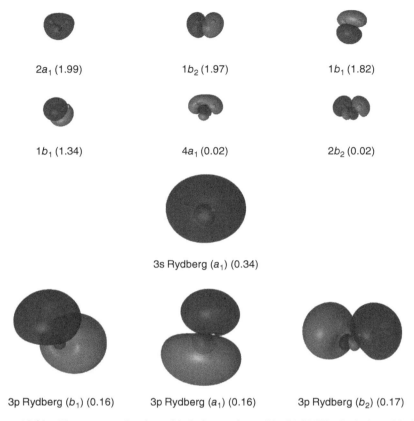

$2a_1$ (1.99) $1b_2$ (1.97) $1b_1$ (1.82)

$1b_1$ (1.34) $4a_1$ (0.02) $2b_2$ (0.02)

3s Rydberg (a_1) (0.34)

3p Rydberg (b_1) (0.16)   3p Rydberg (a_1) (0.16)   3p Rydberg (b_2) (0.17)

Figure 13.20   The converged active orbitals from a 6 root SA-CASSCF calculation of 8 electrons in 10 orbitals in conjunction with an uncontracted molecular Rydberg basis and a triple ζ plus polarization ANO-L basis. The number below the orbitals indicates the SA-CASSCF natural orbital occupation number. Isosurface values for the valence and Rydberg orbitals are 0.040 and 0.017 au, respectively.

the excited states (five states) using the extended active space (including the Rydberg orbitals of interest) and followed by a MS-CASPT2 over the same states. The computed excitation energies compiles to 7.51, 9.26, 9.83, 9.88, and 10.08 eV in conjunction with a quadruple ζ plus polarization ANO-L basis set. These results are to be compared with the results of Rubio and coworkers of 7.50, 9.27, 9.86, 9.95, and 10.15 eV [14]. The minute differences are associated with the fact that Rubio and coworkers selected the way of computing the excited state energies from slightly different SA-CASSCF spaces, in particular states were computed taking symmetry in consideration and only handling states of similar symmetry at each instance.

Summary   In this study of the excitation energies of the lowest state of water, we have experienced some of the special care that has to be considered when making calculations on combined valence and Rydberg state. We have discussed the necessity to use special molecular Rydberg basis set to correctly model the Rydberg orbitals. The creation of the Rydberg orbitals was generated in a calculation on the cationic system. We discussed the selection of the active space orbitals from the context of inclusion of correlating orbitals to get a wave function that has the correct dissociative character and to include orbitals to facilitate the CI to describe the desired excited states. The possible artificial mixing of states of valence and Rydberg character was mentioned and the SA-CASSCF/MS-CASPT2 approach was recommended to resolve such problems.

To conclude, it is worth mentioning that alternative approaches are possible in cases when the valence-Rydberg mixing is artificial in gas phase studies or for absorption spectra in solvents. In the latter case, the solvent effect will blueshift the Rydberg states and the CASSCF active space can now safely be constructed without the inclusion of the Rydberg orbitals. In the former case, if only valence states are of interest, a dirty trick can be performed. The energy of the Rydberg states are much a function of how good the basis set is to describe these states. An artificial blueshift of the Rydberg states can be established by skipping the explicitly designed Rydberg basis and making sure that the remaining atomic basis set is not diffuse enough to describe the Rydberg orbitals appropriately. This will again make it possible to perform the calculation without including the Rydberg orbitals in the active space.

Excited States of para-Nitroaniline

Introduction   The *para*-nitroaniline (pNA) molecule (see Figure 13.21) is an organic chromophore that appears as the intermediate in the synthesis of, for example, dyes and pharmaceuticals. More recently, however, pNA has become of more interest due to it being a simple representative of a "push–pull" class of compounds. This property pNA owes from that the electron accepting nitro group (NO_2) is connected to the electron-donating amine group (NH_2) via a π-conjugated system (the phenyl ring). This results in unique optical properties that make pNA a suitable and simple system for studies of the interaction solvent–solute and how the electronically excited states are affected by the environment [15]. In particular, there is an interest in the solvatochromic shift of the charge transfer (CT) state. This CT state, the result of a photo excitation of an electron from the donor to the acceptor group resulting in a large increase of the dipole moment [16], is sensitive to the solvent and can display a red

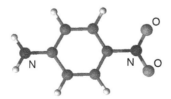

Figure 13.21   The *para*-nitroaniline system, an important precursor in the synthesis of dyes.

solvatochromic shift in polar and polarizable solvents [17, 18]. In this example, we will model this CT state, explicitly taking into account the so-called Rydberg states, as we study states beyond the lowest excited valence state.

The Active Space   The active space for this molecular system will be a challenge due to its complexity. pNA has three different moieties, each alone providing an interesting and rich set of states and in combination even more so. We will find the possibility excited from lone-pairs of the nitro fragment or π orbitals to the π subspace of the benzene fragment. We will have primary amine fragment with a double occupied lone-pair – will this and the out-of-plane π subsystem of the nitro group all conjugate? Some states will be simple valence-excited states, whereas others will be of charge-transfer type; how does this complicate the investigation? In order to answer this question, we will have to design a rather large active space and as we go up in excitation energy, we will have to encounter the Rydberg states too. All these possibilities are not only a challenge to the computational chemist but also forms a litmus test on the quality of the method of investigation.

In as follows the organic procedure of designing the CASSCF/CASPT2 calculation of pNA will be described. The procedure addresses (i) the selection of the valence active space, (ii) the design of a molecular contracted Rydberg basis set, and finally (iii) the tuning of a combined valence and Rydberg excited states calculation. The first part is trivial, or so it might seem. Let us form our initial set of starting orbitals from our standard tool – localized, occupied, and virtual orbitals from an SCF calculations in conjunction with a minimal basis (MB) ANO-L basis set. Considering the energetically accessible orbitals for optically active electronic excitation, we should consider all the out-of-plane π orbitals of the three different fragments and in addition to this we will include the in-plane lone-pairs of the oxygen atoms in the nitro fragment (see Figure 13.22). This now adds up to 16 active electrons in 12 active orbitals (16-in-12). This seemed to be straightforward; however, here we encounter the first problem. The resulting optimized active orbitals did not come out in accordance with our expectations – the in-plane lone-pair orbitals of the oxygen atoms are missing. All the out-of-plane π orbitals are fine; however, the in-plane lone-pair orbitals has been replaced by two orbitals made of the oxygen 2s orbitals (see Figure 13.23). This is a manifestation of the fact that the CASSCF energy expression is rather insensitive to rotations between (i) inactive orbitals and active orbitals with close to double occupation and (ii) virtual orbitals and active orbitals that are close to empty. Cases (i) and (ii) typically are encountered for σ bonds at the equilibrium distance; the bonding orbital is close to double occupation and the antibonding orbital is almost empty. This will change, however, as we stretch the bond. This is clearly not the case here. Case (i) can also be encountered for lone-pairs, obviously the case for this molecular system. To fix this, we need either to increase the active space or even better change the calculation such that the lone-pair orbitals do not have close to double occupation. Since the first excited states of this system is of $n - \pi^*$ character, we achieve this by doing a SA-CASSCF calculation on the ground state and the two lowest excited states. In this way, the SA-CASSCF natural orbital occupation number of the oxygen lone-pair orbitals will not be close to double occupation. Following this procedure, we get the

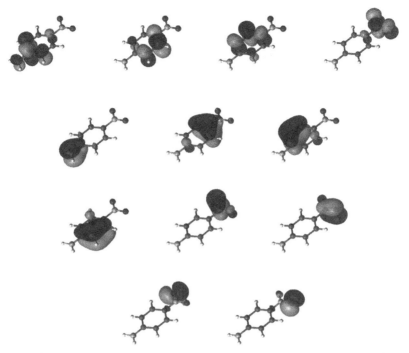

Figure 13.22 The starting active orbitals of *para*-nitroaniline system using a ANO-L minimal valence basis set.

optimized SA-CASSCF active orbitals as follows from Figure 13.24. This concludes the first step in the design of the proper CASSCF/CASPT2 calculation of the optical spectrum of pNA. Note, however, we have not yet included any Rydberg orbitals in the active space and the atomic basis set is still of a very constrained nature.

In the second phase of the study of pNA, we will design a *molecular* contracted Rydberg basis set. The procedure described here does in all central parts follow the procedure as described by Serrano-Andrés et al. [19]. The first step is to compute the position of the charge centroid of pNA$^+$. This is done in a GS 11-in-10 doublet CASSCF calculation, the oxygen lone-pairs are placed in the inactive space and one electron is removed as compared to the SA-CASSCF calculation on the neutral pNA. The valence ANO-L basis set is expanded up to polarized valence double ζ quality. This center is found to be close to the carbon on the bond that joins the phenyl ring with the amine group. In the second step, an uncontracted 8s8p8d Rydberg-type Gaussian basis set, as described by Kaufmann [20], is positioned at the charge centroid in yet another CASSCF 11-in-10 calculation of on pNA$^+$. The three series of Rydberg orbitals, the s-, p-, and d-series, are found as virtual orbitals with a negative orbital energy –a manifestation of their electron affinity. From these orbitals, a series of contracted molecular Rydberg basis functions are generated to be used in

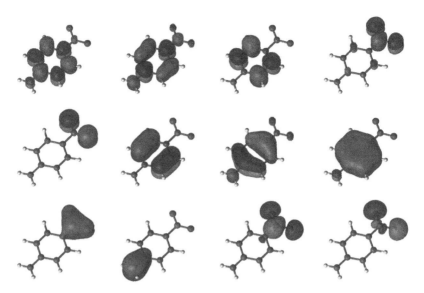

Figure 13.23 The optimized single-root active orbitals of *para*-nitroaniline system using a ANO-L minimal valence basis set. Note that the leftmost orbitals on the lower row correspond to oxygen 2s orbitals and not as expected a pair of in-plane lone-pair orbitals.

the subsequent calculation on the valence and Rydberg lower excited states of pNA (see Ref. [19] for computational details). In the third phase of these calculations we now add the contracted molecular Rydberg basis set, contracted as 8s8p8d to 2s2p2d. To the active "valence" space, we now add the Rydberg set of orbitals. If we like to include the 3s, and 3p Rydberg states we need to add nine active orbitals. This would reduce to 16 electrons in 16 orbitals CASSCF. This would, however, yield a too large a CI expansion and for that reason, an RASSCF/RASPT2 approach was considered. In these calculations, we moved the two oxygen lone-pairs to the RAS1 active space; the RAS1 space was comprised of the out-of-plane π-system and finally the RAS3 space was selected to contain the Rydberg 3s and 3p orbitals. The RAS1 and RAS3 active spaces were constrained to contain at most one hole and one particle, respectively. The ability of these types of RASSCF/RASPT2 models to reproduce the result of the CASSCF/CASPT2 parent model with no significant loss of accuracy have been documented by Sauri et al. [21]. This leaves us with just one final parameter to adjust – the number of roots in the SA-RASSCF. It might be tempting to only select the lowest roots corresponding to the energy range of interest. However, this is to oversimplify the situation. As we know the order of the states will change as we include the dynamical correlation. Furthermore, we might need to extend the number of CI roots beyond the trivial if we like to stabilize a particular active space. In the case of pNA and the suggested RASSCF active space suggested above, we found a reasonable stability first after including 16 roots in the SA-RASSCF (see Figure 13.25 for the associated natural active orbitals). This is still not perfect – note that the last RAS3 orbital is not

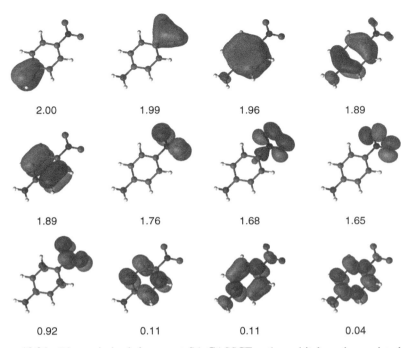

2.00   1.99   1.96   1.89

1.89   1.76   1.68   1.65

0.92   0.11   0.11   0.04

Figure 13.24   The optimized three root SA-CASSCF active orbitals and associated natural orbital occupation numbers of *para*-nitroaniline system using a ANO-L minimal valence basis set.

of 3p Rydberg character and in comparison with the two other 3p Rydberg orbitals is associated with a rather small occupation number. It is apparent that the missing 3p Rydberg state is not found among the 16 lowest SA-RASSCF states. We could continue and expand the number of roots until it was found. Note, however, that the 15 excited states examined so far are 4.10–7.95 eV above the ground state. Since we are only interested in states up to 6.3 eV (197 nm), we can for all practical purposes stop the expansion to include more roots. To conclude this investigation, the final theoretical vacuum spectrum together with vapor spectra are depicted in Figure 13.26.

Summary   In this example, we have experienced the somewhat elaborate process of designing a combined valence and Rydberg excited states calculations. First, a valence active space has to be designed. The number of SA-CASSCF roots had to be chosen such that the space of selected active orbitals is stable. Note that this number of roots might chance with the quality of the valence basis set since the basis set can effect the order of the states. Second, a molecular Rydberg basis set was designed from calculation on the associated cationic system in the ground state. Third, the combined valence and Rydberg states are explored. In some cases, this will be a too large a calculation and an RASSCF/RASPT2 approach has to be used. Again the number of CI roots has to be varied in order to ensure that the selected active space is stable.

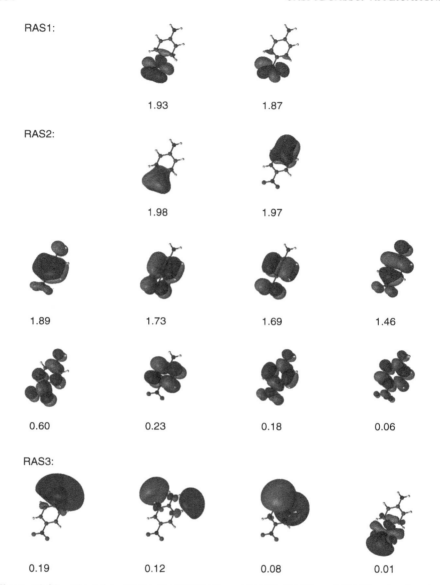

Figure 13.25 The 16-in-16 SA-RASSCF 16 root RAS1, RAS2, and RAS3 natural active orbitals and associated occupation numbers of p-nitroaniline in conjunction with an ANO-L VDZP basis set. Isosurface at 0.02 au.

13.2.3 Photochemistry and Photophysics

In the few following examples, we will study photophysical and photochemical processes. These studies are more complicated than ground state chemistry; in that additional orbitals has to be included in the active space for the excited states and

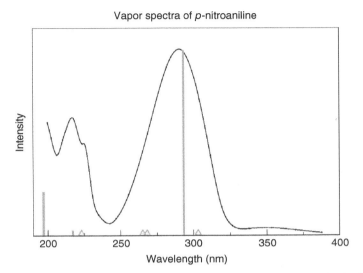

Figure 13.26   The experimental vapor spectra of *para*-nitroaniline by Millefiori [18] and the computed RASPT2 values of the transitions in the 197–350 nm region. The intensity scale is arbitrary. The experimental curve is a digitalization from the 1977 publication. The solid gray lines are the three most intense RASPT2 transitions normalized to have the most intense transition, at 293 nm, to normalize to the same height as the most intense experimental transition. The triangle symbols indicate transitions which, for all practical purposes, can be considered dipole-forbidden transitions.

their chemistry and physics to be modeled correctly. In the following sections, we in particular look at the thermal dissociation of 1,2-dioxetane – a photochemical process; the photostability of uracil – a photophysical process; and finally the pyridine–cyanonaphthalene system – an exciplex.

The Thermal Dissociation of 1,2-Dioxetane

Introduction   The molecular orbitals of 1,2-dioxetane (see Figure 13.27a) provides a good example of chemiluminescent phenomena. The 1,2-dioxetane system will undergo a thermal dissociation, in which, along the dominant reaction path, two ground state formaldehyde molecules are formed. This is not, however, the chemiluminescent path. In the chemiluminescent reaction path, one of the product formaldehyde molecules is formed in an excited state and will relax to the ground state electron configuration emitting light – chemiluminescence. The chemiluminescent path is an example of a nonadiabatic process in which an excited state is populated via a conical intersection close to the transition state (TS) of the O–O bond breakage (see Figure 13.27b) [22]. In reality the conical intersection is not a point but rather a 3N-eight-dimensional manifold, often called a "seam" that joins two different electronic states. In 1,2-dioxetane, this seam starts right after the TS following

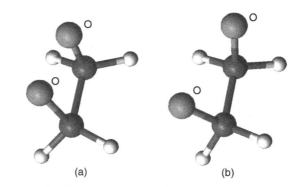

Figure 13.27   The molecular structure of 1,2-dioxetane at (a) the ground state equilibrium structure and (b) the O–O bond-breaking transition state.

the oxygen–oxygen dissociation and extends to where the 2nd TS, the carbon–carbon dissociation, takes place. To further complicate the picture, in the Born–Oppenheimer model, this seam is also an interstate (singlet-triplet) crossing and an entropic trap [23]. Originally, researchers thought that the process observed for the 1,2-dioxetane was the prototype for the bioluminescence observed in living species. The current understanding of these reactions, however, indicate that there are significant differences between the chemiluminescence observed for 1,2-dioxetane and the bioluminescence found in, for example, the luciferin–luciferase system of the firefly [24].

While the 1,2-dioxetane chemiluminescent process has a low quantum yield, emits from a triplet state and is generated via a biradical mechanism, the luciferin system has a high quantum yield (0.41), emits from a singlet state and is facilitated by a charge-transfer-induced luminescence (CTIL) mechanism. In this example, we will study the thermal dissociation of 1,2-dioxetane.

The Active Space   The dissociation of the 1,2-dioxetane system into two formaldehydes involves the breaking of the oxygen–oxygen and carbon–carbon σ-bonds. To model this, we need to include the σ and σ^* orbitals of both of these bonds. In addition, for the intermediate species, after the O–O bond breakage, but before the C–C bond rupture, the lone-pairs of the oxygens will be semi-indistinguishable to the radical orbitals. It is this degeneracy that is the origin of the conical intersections at the vicinity of the first TS. Hence, we need to add the two lone-pairs of the oxygen atoms. In total, for a qualitative correct model of the thermal dissociation of 1,2-dioxetane, we need an active space including eight electrons in six orbitals. This active space will correctly describe the bond-breakage process and the critical degeneracies experienced after the first TS.

Figures 13.28 and 13.29 display the active orbitals and the corresponding occupation numbers of the ground state and the lowest excited state at the equilibrium structure and the O–O transition state, respectively. First, we note that the C–C σ bonds are almost unaffected by the structural difference between the ground or the lowest excited state. For the ground state, we note a significant increase of the occupation of

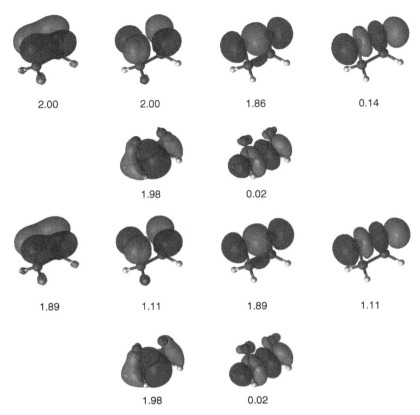

2.00   2.00   1.86   0.14

1.98   0.02

1.89   1.11   1.89   1.11

1.98   0.02

Figure 13.28 The state-specific SA-CASSCF natural orbitals and the associated occupation numbers for the ground state (top two rows) and the first excited state (bottom two rows) at the ground state equilibrium structure of 1,2-dioxetane. Isosurface at 0.04 au.

the σ^* orbital at the transition state as compared to the equilibrium structure (0.71 vs 0.14), reflecting that the energetical difference between the bonding and the antibonding O–O σ bonds is reduced. This is also expressed in the CI expansion in which two electronic structures compete. The first one is of course the HF configuration, with double occupations of the two lone-pair orbitals and the two σ bonds. The second most important electronic structure at the TS is no surprise –a double excitation from the O–O bonding to the O–O antibonding σ orbital. The excited state corresponds approximately to an electron excitation from the antisymmetric combination of the oxygen lone-pair orbitals to the antibonding O–O σ bond. A second electronic structure is, however, also significant and corresponds to a similar excitation, but now taking place between the corresponding symmetric combinations of lone-pairs and the bonding O–O σ orbital. Furthermore, we note that for the excited state, at both structures, an apparent redundancy exists between the oxygen 2p orbitals perpendicular to the C–O bond. For example, at the equilibrium structure

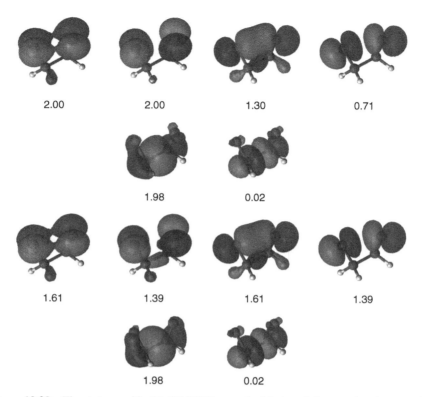

Figure 13.29 The state-specific SA-CASSCF natural orbitals and the associated occupation numbers for the ground state (top two rows) and the first excited state (bottom two rows) at the transition state of the O–O bond breaking in 1,2-dioxetane. Isosurface at 0.04 au.

(see Figure 13.28) the symmetric combinations of the lone-pair orbitals and the corresponding symmetric combination of the 2p oxygen orbitals making up the O–O σ bond have a degenerate occupation of 1.89 in the lowest excited state. For the anti-symmetric combinations of the same orbitals the degenerate occupation is 1.11.

Summary   In the thermal dissociation of 1,2-dioxetane, the product is two formaldehyde molecules. To correctly describe this process, the active space must include both the bonding and antibonding orbitals of the O–O and C–C. Furthermore, the oxygen lone-pairs must also be included in the active space in order to describe the excited states that are involved in the nonadiabatic process. These orbitals are also required to describe the lowest excited state of the formaldehyde fragment –a $n \rightarrow \sigma^*$ state.

Uracil Photorelaxation

Introduction   The uracil molecule, a pyrimidine (see Figure 13.30) is, together with thymine, cytosine, adenine, and guanine, the five natural DNA/RNA monomers. They

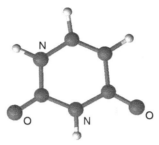

Figure 13.30   The molecular structure of uracil, 2,4-dioxopyrimidine, one of the five DNA/RNA bases.

are distinguished from many other possible purine and pyrimidine-derived systems by having special photochemical properties that protect them from UV damage [25]. The response of these systems to damaging sun radiation (in the optical and UV range) is designed to effectively dissipate absorbed radiation and produce a "hot" photophysical product. In this respect, the response of the system to radiation is photophysical rather than photochemical.

The Active Space   The study of the photostability of uracil requires that we can model the ground state, the initial excited state (the state with the largest absorption probability – we expect this to be a low-lying charge-transfer excited state) and the system along the minimum energy path as it relaxes due to interaction with the environment. In this photorelaxation process, the system will descend to lower excited states via either weakly avoided crossings or conical intersections. This process will be radiationless and to be purely photophysical, it will require that the passage to lower state is unidirectional – transitions from intermediates (at weakly avoided crossings) can result in two different products and passages through conical intersections can potentially result in several different products. With this in mind, we design the active space from out-of-plane p-orbitals of the carbon, oxygen, and nitrogen atoms, which will make up the set of π and π^* orbitals. In addition, we include the lone-pair orbitals of the oxygens. This is based on the assumptions that the lowest excited state with the highest absorption probability is most likely a $\pi \leftarrow \pi^*$ state, and that the lowest excited state is a $n \leftarrow \pi^*$ state. The active space will then have 10 orbitals with 14 electrons, as shown in Figure 13.31.

   In the attached figure (see Figure 13.32), we depict the energy of the three lowest states – the ground state, the dark $n \leftarrow \pi^*$ state, and the $\pi \leftarrow \pi^*$ state. The reaction coordinate starts from the Frank–Condon structure being positioned on potential energy surface of the $\pi \leftarrow \pi^*$ state, a state with the highest oscillator strength for a transition from the ground state. First, there is a short time relaxation of the molecular structure associated with a large decrease in energy. This is followed by slower dynamics in which the molecular system progress on the moderately repulsive to flat part of the energy surface. The dynamics then proceeds to and through the first of two conical intersections or weakly avoided crossings. The first such event is where

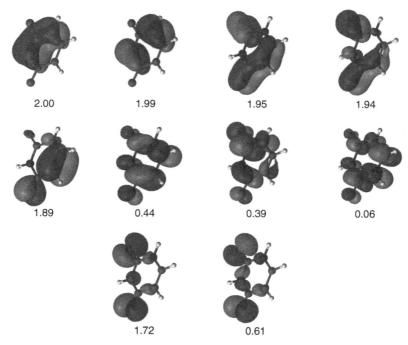

2.00   1.99   1.95   1.94

1.89   0.44   0.39   0.06

1.72   0.61

Figure 13.31 The SA-CASSCF natural orbitals and the associated occupation numbers for the uracil molecule with a valence triple zeta plus polarization basis set. Top two rows depict the eight π orbitals and the lower row the oxygen lone-pairs. Isosurface at 0.04 au.

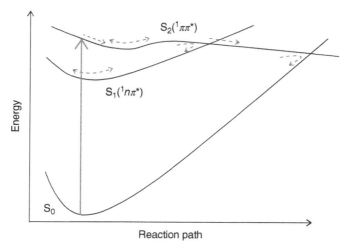

Figure 13.32 Cartoon showing the excited state dynamics in uracil after UV absorption. *Source:* Matsika 2013 [26] Reprinted with permission of American Chemical Society.

the $\pi \leftarrow \pi^*$ and $n \leftarrow \pi^*$ become degenerate. The wave packet here will split. It can either proceed with a contraction back toward the equilibrium structure on the $n \leftarrow \pi^*$ surface, or forward on the $\pi \leftarrow \pi^*$ state potential energy surface. After the conical intersection has been passed, this becomes the first excited state. The second conical intersection is found when the ground and the $\pi \leftarrow \pi^*$ states become degenerate. Here the wave packet will in a nonradiative process enter the ground state surface and start the relaxation toward the ground state equilibrium structure, completing the path in which the nonradiative process converts the radiative energy into thermal energy represented by a hot species on the ground state surface.

Summary In this section, we have briefly discussed the selection of the active space of uracil for the purpose of simulating the photochemistry and photophysical processes subsequent to a photo excitation in the UVA and UVB energy range. The consideration of both the lowest states and the photo active states – dark and bright states – is essential for a proper modeling. The selection of the active space of uracil is in no way more complicated as compared to the other nucleobases: adenine, thymine, cytosine, and guanine. Similar rules and considerations should also be applied to molecular systems with conjugated π systems and oxygen lone-pairs.

Pyridine–Cyanonaphthalene

Introduction It has recently been suggested that a "bonded complex" is formed in the first excited state of the complex formed between pyridine and a number of cyanoaromatics [27]. One of the simplest examples where formation of the exciplex takes place is an interaction between pyridine and cyanonaphthalene.

The interaction between pyridine and cyanonaphthalene, if we restrict ourselves by considering only the ground state, is repulsive. It means that any calculation, performed within Hartree–Fock or DFT approach, will predict that such complex does not exist. In order to describe the creation of the complex, we have to study the excited states for this system. On the other hand, the multiconfigurational treatment of the complex seems to be problematic: the total number of unsaturated orbitals plus lone pairs is 23, which is far beyond the possibilities of CASSCF approach.

We may try to describe pyridine–cyanonaphthalene complex by a limited number of active orbitals (keeping in mind that the different set of orbitals can be important for description of isolated molecules and the complex), or we can explore the possibilities of a RASSCF approach [28].

The Active Space We choose the ANO-S-VDZP basis set (yielding in total 312 basis functions), and use geometries optimized by the DFT approach. The pyridine molecule is placed above one of the carbon atoms of cyanonaphthalene (Figure 13.33), producing a small distortion in cyanonaphthalene geometry: the hydrogen atom attached to a carbon facing the pyridine nitrogen atom gets displaced out-of-plane when the two molecules get closer. This distortion of the monomers was small, according to the DFT calculations, and could be ignored in a qualitative description of this system. Thus, the RASSCF calculation could be done using only the relative degrees of freedom between the two rigid monomers.

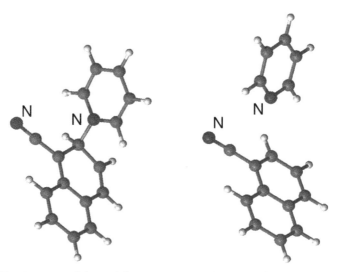

Figure 13.33 Structure of the pyridine–cyanonaphthalene complex at equilibrium distance 1.45 Å and at 2.8 Å.

In the first study, we selected 10 active orbitals at the CASSCF level, intended to have 8 orbitals located on cyanonaphthalene and only two on pyridine (the lone-pair and one extra correlating orbital), leaving the pyridine π system inactive.

This turned out to be a serious mistake: The pyridine π-orbitals play an important role in the lowest excited state, the charge transfer state. The calculations were therefore performed with 14 active orbitals (Figure 13.34).

This is a very large calculation with almost 6 million determinants in the wave function. It is difficult to know if this choice is sufficient, since we could not at that time increase the active space any further. However, the comparison between calculations with 12 and 14 active orbitals indicates that 14 active orbitals might be sufficient for the description of this system.

The calculations confirm that the ground state is repulsive, with a small kink around 1.5 Å. The first excited state has a local minimum at 1.47 Å with excitation energy 1.73 eV, which correlates with experimentally observed data.

By limiting the excitation levels, one can in this case use a much larger active space. The active space is subdivided into three spaces: RAS1, RAS2, and RAS3. The RAS1 space is in principle doubly occupied, but a certain maximum number of electrons may be excited to RAS2 or RAS3. In these calculations, up to double excitations are allowed. The RAS2 space corresponds to the old CAS space and all possible types of occupations are allowed. Finally, the RAS3 space is in principle empty but is allowed to accept up to a maximum number of electrons. In our calculations, we allowed at most two electrons in the RAS3 orbital space.

By these restrictions, we could include all 23 orbitals. From the CASSCF calculations, it was found that most orbitals had occupation number close to two or zero,

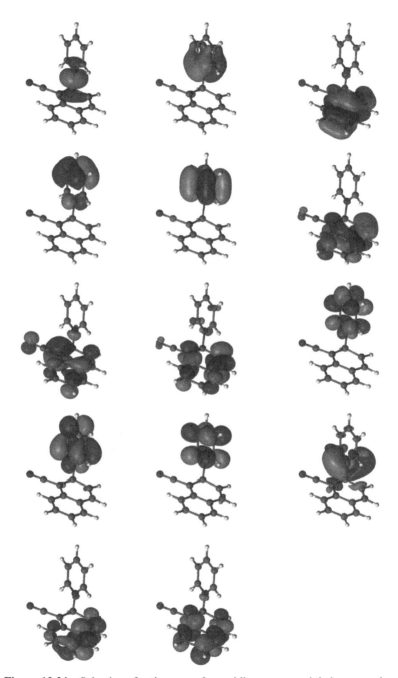

Figure 13.34   Selection of active space for pyridine–cyanonaphthalene complex.

with three prominent exceptions: three natural orbitals had occupation numbers 1.33, 0.40, and 0.37. These were then used as RAS2 orbitals, and we denote this RAS space as /11/3/9/. All the orbitals of the two π system, plus the CN in plane π orbitals and the orbitals of the N(pyridine)–C(cyanonaphthalene) bond are active. For some reason, the system wants to have four orbitals (two bonding and two antibonding) in this region.

Thus, there are now no orbitals missing in this treatment. The price paid is that the restrictions in the excitation level may be insufficient. However, the results obtained around equilibrium are very similar to those we obtained in the CASSCF calculations. The potential energy curves are almost identical. The equilibrium distance is now computed to be 1.472 Å, compared to 1.473 Å in the CASSCF calculation. The excitation energy is somewhat lower, 1.43 eV compared to 1.73 eV. The fact that the two calculations give similar results is gratifying and confirms that we are getting a correct picture of the situation around equilibrium.

Using the RASSCF method with small RAS2 space allows us to include all π orbitals into active space, without visible changes in the results. As a side effect, RAS calculation is much faster, as compared to the 14-orbital CASSCF calculation, and leaving an open question about the optimal way of selection of these orbitals.

Although the active space with only three orbitals in RAS2 space seems to be sufficient for description of the pyridine–cyanonaphthalene complex with a short separation between molecules, we should not expect that this active space will be sufficient for all interatomic distances. In the case of a large separation between molecules, the excitation energy is related to pyridine molecule, which is not well described by our selection of the active space.

Summary A weak bond between two molecules, which appears only for an excited state is, of course, rather unusual example that demonstrates that the trends obtained for the ground state and exited states can be different. We also have to note that neither density functional theory, or time-dependent DFT (which is applicable only in the cases where the excited state resemble the ground state) will be able to describe the formation of this complex.

At the same time, the presented system clearly demonstrates the limitations of a multiconfigurational approach for description of even moderately sized organic molecules. If a system contains a large number of nonsaturated bonds, aromatic rings and unpaired electrons, the selection of the active space becomes a nontrivial but essential step in the calculation. Using compromise solutions with approximate treatment of excitations, such as RASSCF/RASPT2, may be the only affordable option. On the other hand, new methods, such as DMRG, may in the future offer less costly ways of including large active spaces.

13.2.4 Transition Metal Chemistry

In this section, we view two examples involving the presence of transition metals – in chloroiron corrole and in a metal–metal bond of transition metals. In the latter case, the concept of effective bond order will be the primary subject of discussion. First, however, we will have some general notes on the complications that transition metals offer in the process of selecting an active space for a CASSCF calculation.

Transition Metals and the Double d-Shell Effect

Introduction   Calculations involving transition elements must often take account of the *double d-shell* effect. This refers to the need to include additional correlating orbitals of d type, as was discovered by Froese-Fischer: the $d^{10}(^1S)$ state of the Ni atom [29] was better described as a d^9d' state, where the d' orbital is more diffuse than the d-orbitals. This was not completely unexpected; for example, some of earliest He atom calculations used wave functions essentially of the type ss' for the radial correlation, but what was surprising was the size of this effect. It was soon clear that this effect had to be accounted for in many calculations involving first-row transition elements.

Electron-rich bonds, as well as electron-rich open shells in atoms, can give rise to correlation effects that are neither of the clear-cut open-shell nature of traditional multiplet coupling nor of the simple "near-encounter avoidance" type that is well accounted for by a DFT correlation functional. This may be rationalized in terms of a coupling between the motion of valence electrons and the nonlinear response of a background charge cloud containing several or many electrons. Strong such effects are present in the valence region of some of the transition metals, the lanthanides, and in multiple bonds with effective bond order lower than the formal bond order (partially broken multiple bonds).

Returning to the Ni atom, early CASSCF/CASPT2 calculations by Andersson and Roos [30] showed that the natural occupation number of the correlating d' orbital, which is a measure of this double-shell effect, depends on the electronic structure of the atom. In CASSCF wave functions that includes the second d-shell in the active space, the sum of the natural occupation numbers over the d and d' shells is very close to the numbers expected from conventional multiplet theory, namely 7.99 for the d^8s^2 (^3F) state, 8.98 for the d^9s (^3D, ^1D), and 9.97 for the $d^{10}(^1S)$ state, while the corresponding number for d' alone increased strongly with this d population, from $0.04(^3F)$ via $0.07(^3D,^1D)$, to 0.12 (^1S). Qualitatively, the more crowded the d-shell, the stronger is the correlation, which results in an approximate picture such as

$$\Psi = \left(\hat{1} + \sum_{\sigma \in \{\alpha,\beta\}} \sum_{m=-2}^{2} \hat{b}_{m\sigma}^{\dagger} \hat{a}_{m\sigma} \right) \Psi_0,$$

where in this formula $\hat{b}$ and $\hat{a}$ would correspond to orthonormal d' and d orbitals, respectively, and where Ψ_0, would be a conventional LS-coupled multiplet state of the OSHF type where just the d orbitals are occupied.

This is what an atomic CASSCF wave function could look like for a radial $d^{n-1}d'$ configuration, properly symmetrized. However, the active orbitals can be transformed among themselves at will, and get (almost uniquely) defined only by requiring them to be natural orbitals. In a CASSCF calculation using orthonormal natural orbitals, the single excitations disappear, and the customary double excitations are seen instead.

The Active Space of First-Row Transition Metals and their Organic Complexes A good active space for a late transition element thus needs two complete d-shells and the s-shell. For accuracy, also the p-shell should be correlated. Early first-row, and second-row, transition elements have less need of a double d-shell. Also, the need for the additional d-shell in the active space is largest when each d-orbital appears with varying occupation. When formally counting the number of orbitals to have active, it is easiest, and sufficient, to count the d-shell as requiring 10 d-orbitals; the charge-separating correlation of a bond to a ligand does *not* require an extra orbital, since those d-orbitals that are engaged in covalent bonds are no longer free to have a (strongly) fluctuating shape. If the number of active orbitals is too large, strong, and permanent, metal-ligand bonds may be treated as inactive. High spin obviously helps, by eliminating the possibility for fluctuating occupation.

For molecules with more than one transition element, the double-shell rule cannot be followed in a conventional CASSCF calculation, since already the d-shells would require 20 orbitals. Too severe compromises can be avoided by using RASSCF, but at the price of slower convergence, and increased risk of false convergence, or even no convergence, and, of course, size inextensivity. False convergence occurs very often through a (strong) breaking of (local) symmetry and may require experimentation with, for example, how many states to include in a state-average optimization. Problems can be avoided by preliminary calculations with small basis sets.

Chloroiron Corrole

Introduction Transition metal corroles may be viewed as stable synthetic analogues of high-valent heme protein intermediates.

The spin and the actual valence state of the transition metal in iron corroles is not known. NMR spectroscopy indicates that chloroiron corrole is best described as Fe(III) centers (with $S = 3/2$) coupled to corrole^{2-} radicals. Nevertheless, the description of chloroiron corroles (see Figure 13.35) that includes high valence Fe(IV) is also popular. Calculations, performed within DFT, cannot conclusively support or reject the hypothesis about high valence iron and its spin.

To resolve the debate, one should make multiconfigurational study of the system and compare the energy of the states [31].

The Active Space If we try to apply the rules for selection of the active space – we will get a size of the active space that is clearly much larger than any computational code can handle. The selection of the active space for a system such as chloroiron corrole requires compromises and experimenting.

Calculations were performed with a fixed geometry. The symmetry (C_s) has been used in this study. The relativistic ANO-RCC basis set has been chosen. For Fe, Cl, and N, we used VTZP quality (which corresponds to 6s5p3d2f1g for Fe) and VDZP basis set has been used for C and H. In total, it corresponds to 534 basis functions.

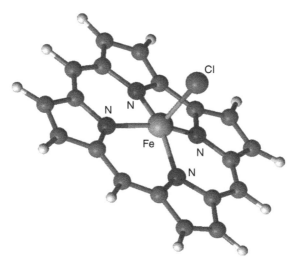

Figure 13.35 The chloroiron corrole molecule.

A large number of preliminary studies have been performed using different active spaces, which all had one or another defect. We can present these attempts in "historical" order, explaining a usual logic behind the process of the selection of the active space.

1. The first idea was to include the 3d orbitals of iron together with orbitals describing the nitrogen and chlorine ligands. This resulted in an active space of 16 electrons in 11 RAS2 orbitals (16in11). This active space captured the main features of the electronic structure, including the radical character of the lower states with one electron in a ring orbital. But it was missing a number of important features. It included no ring orbitals, except the one involved in the radical pair. Instead it included a number of orbitals with occupation numbers very close to two, which could equally well have been left inactive.

2. In another attempt, the active space was extended such that more ring orbitals were included at the expense of the doubly occupied nitrogen and chlorine lone-pairs. In order to have the same active orbitals for both states, it was necessary to extend the active space to 16in14, and use the symmetry to partition the active space. The calculations showed that all the orbitals included are important in the sense that they have occupation numbers between 0.03 and 1.97 in at least one of the states.

3. The previous active space still has a serious flaw. One of the 3d orbitals is inactive. Thus, it has to be added and this gives an active space of 18in15, which is getting close to the limit. Calculations with this active space can only be made for one root in each symmetry.

4. Even if that calculation was successful, this active space was still not optimal. It had a number of ring orbitals active but lacked correlating orbitals for 3d

(4d, the double-shell effect). It was therefore decided to remove six of the ring orbitals and the corresponding electrons and instead add two carefully chosen 4d orbitals. This gave active space 12in11. The results were, however, not very encouraging. Different types of active orbitals were obtained in the different symmetries and spins. Some states kept the 4d orbitals active, but others replaced one of them with an antibonding ring orbital and also shuffled an inactive ring orbital into the active space.

5. Detailed analysis of orbitals showed that the problem in selection of the active space for chloroiron corrole is related to the fact that for different roots, different orbitals need to be included. An active space that can be considered as a compromise solution includes 14in13.

   This active space gave a consistent result for the lowest septet, quintet, triplet, and singlet states in both symmetries (eight states). The same orbitals described all states. All five 3d orbitals and two 4d orbitals are active together with the ligand orbitals that interact with the iron orbitals. In addition, there is one pair of ring orbitals (one doubly occupied and the corresponding correlating orbital), which the molecule decided that it wanted to have active. We are therefore confident that this active space will describe the electronic structure well and be able to discriminate between the different low lying electronic states.

6. The above 14in13 active space seems to cover the essentials of the low lying electronic states. However, this active space is unbalanced due to the lack of ring orbital of a'' symmetry that can couple to the $3d_{xz}$ orbital. Adding one orbital in symmetry a'' gives more balanced active space for $^3A'$ and $^5A''$ states.

The final orbitals and associated occupation numbers for $^3A''$ state are presented in Figure 13.36.

Relative energies (eV), computed in CASPT2 level for lowest singlet, triplet and quintet states in chloroiron corrole are shown in Figure 13.37.

Calculations confirm multiconfigurational character of the electronic structure of chloroiron corrole. There is no evidence of a Fe(IV) state, high- or low-spin, within 1.5 eV of the ground state.

As we can now conclude from the table, the difference between 14in13 and 14in14 calculations affects only higher excited states. Active space 14in14 seems to be a reasonable choice for the description of chloroiron corrole.

Now, knowing the answer (which comes as a result of many try-and-fail attempts), we can ask ourselves: is it possible to achieve the same conclusion in more straightforward or, at least, in semiautomatic way?

The size of the active space is already large enough, so it would be difficult to increase it. However, we can try to use a restricted method such as RASSCF, which treats excitations in an approximate way. The corresponding cost is substantially cheaper, and we can include into RAS1/RAS3 active space a much larger number of orbitals. For example, for our system 13 active orbitals (in RAS2 space), split by symmetry into subset of seven and six orbitals produce 613 320 determinants. Using 33 active orbitals in RAS1/RAS3 active space with allowing only single and double

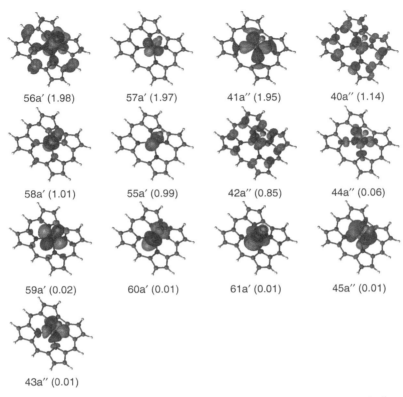

56a′ (1.98) 57a′ (1.97) 41a″ (1.95) 40a″ (1.14)

58a′ (1.01) 55a′ (0.99) 42a″ (0.85) 44a″ (0.06)

59a′ (0.02) 60a′ (0.01) 61a′ (0.01) 45a″ (0.01)

43a″ (0.01)

Figure 13.36 The final orbitals and associated occupation numbers for the $^3\text{A}''$ state of chloroiron corrole as computed with a 14in13 CASSCF.

State	14in13	14in14	Occupation
1 $^3\text{A}''$	0.00	0.00	$(3\text{d}_{x^2-y^2})^2(3\text{d}_{yz})^\uparrow(3\text{d}_{z^2})^\uparrow(3\text{d}_{xy})^0(3\text{d}_{xz})^\uparrow(a')^\downarrow$
1 $^5\text{A}'$	0.37	0.33	$(3\text{d}_{x^2-y^2})^\uparrow(3\text{d}_{yz})^\uparrow(3\text{d}_{z^2})^\uparrow(3\text{d}_{xy})^\uparrow(3\text{d}_{xz})^\uparrow(a')^\downarrow$
1 $^5\text{A}''$	0.56	0.54	$(3\text{d}_{x^2-y^2})^2(3\text{d}_{yz})^\uparrow(3\text{d}_{z^2})^\uparrow(3\text{d}_{xy})^0(3\text{d}_{xz})^\uparrow(a')^\uparrow$
1 $^7\text{A}'$	0.67	0.62	$(3\text{d}_{x^2-y^2})^\uparrow(3\text{d}_{yz})^\uparrow(3\text{d}_{z^2})^\uparrow(3\text{d}_{xy})^\uparrow(3\text{d}_{xz})^\uparrow(a')^\uparrow$
1 $^3\text{A}'$	0.82	1.13	$(3\text{d}_{x^2-y^2})^2(3\text{d}_{yz})^\uparrow(3\text{d}_{z^2})^\uparrow(3\text{d}_{xy})^0(3\text{d}_{xz})^\uparrow(a'')^\downarrow$
1 $^7\text{A}''$	1.29	1.23	$(3\text{d}_{x^2-y^2})^\uparrow(3\text{d}_{yz})^\uparrow(3\text{d}_{z^2})^\uparrow(3\text{d}_{xy})^\uparrow(3\text{d}_{xz})^\uparrow(a')^\uparrow$
2 $^5\text{A}''$		1.35	$(3\text{d}_{x^2-y^2})^\uparrow(3\text{d}_{yz})^2(3\text{d}_{z^2})^\uparrow(3\text{d}_{xy})^0(3\text{d}_{xz})^\uparrow(a')^\uparrow$
2 $^5\text{A}'$		1.66	$(3\text{d}_{x^2-y^2})^2(3\text{d}_{yz})^0(3\text{d}_{z^2})^\uparrow(3\text{d}_{xy})^\uparrow(3\text{d}_{xz})^\uparrow(a')^\uparrow$
1 $^1\text{A}'$	1.76	1.73	$(3\text{d}_{x^2-y^2})^2(3\text{d}_{yz})^2(3\text{d}_{z^2})^0(3\text{d}_{xy})^0(3\text{d}_{xz})^\uparrow(a'')^\downarrow$
1 $^1\text{A}''$	1.75	1.75	$(3\text{d}_{x^2-y^2})^2(3\text{d}_{yz})^2(3\text{d}_{z^2})^0(3\text{d}_{xy})^0(3\text{d}_{xz})^\uparrow(a')^\downarrow$

Figure 13.37 Relative energies (eV), computed in CASPT2 level for lowest singlet, triplet and quintet states in chloroiron corrole. The symbols ↑ and ↓ indicate up- and down-spin-orbitals. The a' amd a'' orbitals are corrole π HOMOs.

excitations, and using same symmetry (nine and six orbitals in RAS1, and nine and nine orbitals in RAS3), produce only 34 317 determinants.

Thus, we can set up a calculation with known excess of active orbitals, treated in an approximate way, and analyze the obtained result. If an occupation number for an orbital will remain close to pure 2.00 or 0.00, they can be excluded from the active space. And those orbitals that will obtain some intermediate occupation should be treated more accurately.

Ten out of fifteen orbitals included into RAS1 space still have occupation number close to 2.00. Most of these orbitals have π character, and they are localized at corrole rings. Note, that just looking at the shape and spatial localization of these orbitals (see Figure 13.38), we cannot be certain that they might be excluded from the active space.

In Reference [28], chloroiron corrole has been used to check a semiautomatic procedure of selection of the active space, using a set of cheap RASSCF calculations in order to spot the orbitals with occupation numbers deviating from unperturbed 2.0 and 0.0 values. Starting from 33 orbitals, which we selected without having deep thoughts, we can identify orbitals with intermediate occupation numbers. This set is identical to one that was found by try-and-fail procedure. The procedure of "RAS probing" (including large amount of RAS1 and RAS3 orbitals, followed by analysis of occupation numbers) can be useful in case of complex systems with many candidates for the active space.

Summary Selection of an optimal active space is always a compromise. If we apply the formal rules for the selection of the active space to a molecule, which contains one or several transition metals, and ligands with aromatic rings and unsaturated bonds, we will find out that the size of active space is much larger than any modern quantum chemical software can handle, using conventional GUGA CI technology. However, not all excitations are equally important for the proper description of such system.

Using RASSCF/RASPT2 technique can be very helpful in this case, since more orbitals can be included (in an approximate way) into active space.

Metal–Metal Bonds: Bond Order versus Effective Bond Order The bond orders and bond strengths of some metal–metal bonds are analyzed in an interesting article in Ref. [32]. It is often taken for granted that multiple bonds are much stronger than single bonds, although of course it is also well known that each additional bond order adds incrementally smaller strength to the bonds.

The article shows, by a few examples, that what counts is not so much the bond order, but rather *effective* bond order [32] (effective bond order (EBO), see below); but also how a lower bond order may well be associated with stronger/shorter bonds due to interaction with ligands. Ligands can also compress a bond, and La Macchia et al. [33] have shown an extreme example.

A clear-cut example of a hextuple bond is found only in Mo_2 and W_2. Consider Figure 13.39, showing the natural occupation numbers of W_2, and the active orbitals near the equilibrium distance.

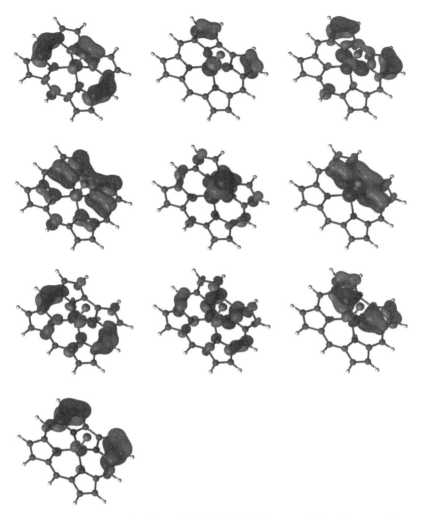

Figure 13.38 The 10 active orbitals in the RAS1 space of chloroiron corrole.

The calculation is a simple one: the valence (5d6s) orbitals of each atom give 12 MOs of which 2σ orbitals, a π shell, and a δ shell are almost fully occupied, and an almost empty set of $2\sigma^*$ orbitals, a π^* shell, and a δ^* shell.

The effect of charge separation correlation is obvious. Starting from dissociated atoms and going toward shorter distances, first the 6s atomic orbitals begin to overlap and split into the $6s\sigma$ and $6s\sigma^*$ pair; then similarly the $5d\sigma$, $5d\pi$, and $5d\delta$ components.

The electronic structure is close to a perfect-pairing wave function, as indicated by the striking mirror symmetry of the orbital occupation numbers. Similar molecular orbitals would be formed for several other diatoms involving two transition metal atoms of either the first or the second row. Such molecules would then have the

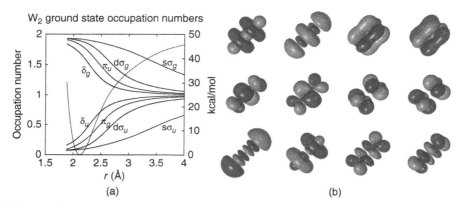

Figure 13.39 The active orbitals of a CASSCF calculation for the W_2 molecule, using 12 active electrons in 12 orbitals. (a) The natural occupation numbers; (b) Orbital pictures.

maximal possible formal bond order six, namely if they contain 12 bonding electrons and no antibonding ones.

However, the contribution from each of the valence MOs to the energy is counterbalanced by the repulsion of the cores, and the precise point where the energy minimum occurs is much dependent on the two species involved. Taking the qualitative nature of the natural occupation numbers as being given, one or more of the short-ranged bonds may be partially broken, depending on where along the curves that the equilibrium bond length is situated. This is indeed the case, and Roos and coworkers defined a more accurate measure of the bond order, which is applicable to most binary bonds: the *effective bond order* [32], which is the sum of the natural occupation numbers of the bonding orbitals minus the sum for the antibonding ones. (The calculation is assumed to be a valence CASSCF, which gives a fairly precise definition, but any correlated treatment that gives a good approximation to true natural orbitals will give similar, if not identical results.) The effective bond orders are of course lower than the formal ones, but in some cases much lower (e.g., Cr_2, EBO = 3.5, but the bond is formally a hextuple bond; Th_2, on the other hand, has a formal bond order of 4, in good agreement with the EBO = 3.7).

It is in fact an almost universal observation that the additional bonds in multiple bonding are partially broken, which makes these prime candidates as active orbitals, and also for both chemical bond activation and photochemical bond breaking. Furthermore, the single excitation from bonding to antibonding orbital has a large transition dipole, and when there are more than one neighboring double or multiple bonds, these excitations couple to give intricate and interesting photochemistry.

13.2.5 Spin-Orbit Chemistry

Spin-orbit coupling is an important effect in chemistry for a number of reasons. A heavy element present in a compound of otherwise light elements has consequences for photochemistry and NMR spectroscopy, the so-called heavy atom effect, but it is generally of importance, as a quick look into Chemical Abstracts or other search services will show. Out of 21 000 hits on the Science Citation Index one-third are from chemistry journals; out of these, the number of hits per year doubled from 2007 to 2014.

For the lighter atoms, the conventional orbitals consists of a single set of spatial orbitals. Each orbital gives rise to a spin-orbital by combining it with a spin function that is either α or β, which stands for an m_s value of $\frac{1}{2}$ and $-\frac{1}{2}$, respectively. In this spin-orbital basis, the operators $\hat{l}^2$, $\hat{l}_z$, $\hat{s}^2$, and $\hat{s}_z$ are diagonal, having the eigenvalues $l(l+1)$, m_l, $\frac{3}{4}$, and $\pm\frac{1}{2}$, respectively. It is then easy to transform to orbitals that are eigenfunctions of the operators $\hat{j}^2$, $\hat{l}^2$, $\hat{j}_z$ instead, with eigenvalues $j(j+1)$, $l(l+1)$, and m_j and indexed using the quantum numbers j, l, and m_j. For a spherically symmetrical mean-field Hamiltonian, the orbitals with given main quantum number n and angular quantum numbers j and l are degenerate and have $(2j+1)$ components. For example, a 3d shell has four degenerate $3d_{3/2}$ components and six $3d_{5/2}$ components, and so on.

For lighter elements, the nlj subshells have almost the same radial extent, and spin-orbit coupling effects nearly cancels for filled subshells. For open nl shells of atoms, the electrostatic interaction is dominating, giving the usual multiplets: While this multiplet splitting depends on spin, this is mainly because different total L and S give rise to different contributions of the usual exchange to the total energy. The spin-orbit interaction is then a smaller effect that splits up the LS term into levels, with energy expressed using a spin-orbit parameter A

$$E(LSJ) \approx E(LS) + \frac{A}{2}(J(J+1) - L(L+1) - S(S+1))$$

provided there is a single open shell. Also molecules may have degenerate configurations with large splitting into electronic terms, which in turn can have smaller spin-orbit splitting. This happens for linear radicals that have orbital momentum, for example, Π or Δ states, but also, for example, in C_3 or other higher symmetries for open shells with symmetry e or t. However, the simplicity of the central mean field is lost in molecules.

For heavy elements (especially for inner shells), the subshells with $j = l \pm \frac{1}{2}$ have quite different extent, and the LS scheme above is no longer accurate.

In polyatomic low-symmetry molecules, the ground state is usually a nondegenerate singlet, and the one-electron levels are each populated by two electrons with $m_s = \pm\frac{1}{2}$. The contribution to the energy from spin-orbit coupling is zero, and one often use the expression that the spin is "quenched." The spin-orbit coupling can begin to gain importance only for open-shell configurations. This is often the case during dissociation to radicals or for excited states, and in such cases the spin-orbit coupling is of importance for reaction rates.

Formally spin-forbidden interactions imply that potential energy surfaces cross without any transition. Even a small spin-orbit interaction between, for example, a triplet and a singlet state will cause transitions. These are important close to a so-called interaction seam, which is the subset of structures where the two states have the same energy. While the interaction itself is small, there is an abrupt change in the electronic wave function, and so a quantitative calculation of such so-called Intersystem Crossing (ISC) rates requires a nonadiabatic treatment. In favorable cases, the ISC rate can be estimated by simple formulas, but usually one would have to use computed nonadiabatic coupling matrix elements.

The catalytic effect of many organometallic compounds is due to spin-orbit interaction, but ISC crossing occurs even for first-row chemistry and ground states, as exemplified by the reaction $^1HNO + OH^- \leftrightarrow ^3NO^- + H_2O$, (in aqua), a spin-forbidden acid/base pair [34]. A recent short review [35] cites computational examples of spin-forbidden reactions.

Relativistic quantum theory is inherently a many-particle quantum field theory, since any number of particle pairs are created or annihilated in the interactions, but for purposes of quantum chemistry, it is sufficient to use models where the electrons interact with nuclei and fields by the direct sum of Dirac one-electron Hamiltonians, one for each electron, and which interact with each other by a Breit–Pauli Hamiltonian. (Neglected effects of, for example, quantum electrodynamics, or nuclear structure, are easily observed spectroscopically, but are insignificant for chemistry [36].) This approximation is thus essentially the same as the usual Schrödinger one, but the orbitals are conventionally written as four-component arrays, each containing four complex amplitudes. However, there are a number of ways to reduce this to two-component form, with insignificant loss of accuracy (see Chapter 5). If magnetic interactions are to be properly described, it turns out that the limiting "nonrelativistic" theory is still two-component (the Pauli equation) rather than the one-component Schrödinger one, which is often assumed. A cross form is to use single-component "scalar" relativity, which is easily applied to any conventional quantum chemical method simply by a redefinition of the usual integrals. When needed, a spin-orbit Hamiltonian can be invoked after the wave function is obtained (see Chapter 10).

The size of the spin-orbit effect varies strongly. As an example, the energy splitting of atomic $P_{1/2}$ and $P_{3/2}$ states is 0.5, 9, and 19 kcal/mol for Cl, Br, and I, respectively. For Br and I, this is too large to be ignored.

The following example shows calculations on electronic ground and excited states of methyl bromide, CH_3Br, for the bromide abstraction.

The Alkyl Bromide Dissociation

Introduction This study explores the ground and excited states of alkyl bromide along the fragmentation process $CH_3Br \rightarrow CH_3 + Br$, where we note that both product fragments are radicals. There are two significant features: First, during the fragmentation process, the carbon atom changes its hybridization from sp^3 to sp^2, going from a tetrahedral coordination to a planar triangular structure. Second, spin-orbit coupling gives two possible final states for the bromine atom. The spin is $S = \frac{1}{2}$, and the orbital quantum number is $L = 1$. By the rules for spin-orbit coupling, the possible J values are then

$$J = S + L, S + L - 1, \ldots, |S - L| \qquad (13.1)$$

so the possible final states for the bromine atom are denoted $^2P_{1/2}$ and $^2P_{3/2}$. According to Hund's rules, the latter is energetically the lowest and is found about 9 kcal/mol below the $P_{1/2}$ state.

Active Space   The active space is then rather simple, we will break the C-Br σ bond. This calls for the active space to include the σ and σ^* C-Br bond. In addition, at dissociation, we have to take care of the degeneracy of the bromine atom – the configuration is $[Ar]4s^2 3d^{10} 4p^5$, with an open 4p shell, having three degenerate (spatial) orbitals (x, y, and z). The whole 4p shell should be in the active space.

The conventional and accepted picture of the photochemistry of the breaking of a σ bond will give us four different electronic states – two diradical states (^1D and ^3D) and two zwitterionic states (Z_1 and Z_2). However, the two zwitterionic states are high energy states and should for practical reasons be avoided here. In particular, they will energetically be in the same range as the Rydberg states. Hence, for the singlet states we will only include the ^1D state and the two states that correspond to ($n \rightarrow \sigma^*$) states. In total, three singlet states are included, and for each singlet state there is a corresponding triplet state to be considered.

For the reactant and product structures, we will use bond lengths $R(CH_3–Br) =$ 1.9599 and 4.00 Å, respectively, and study the active orbitals. Figure 13.40 presents the natural orbitals of the singlet ground state and one of the degenerate $n \rightarrow \pi^*$ states. In particular, we note that the ground state CASSCF wave function contains a small component of antibonding – the natural occupation number of the π^* orbital is 0.03. Thus, already at equilibrium, the ground state CI expansion includes configurations of antibonding character.

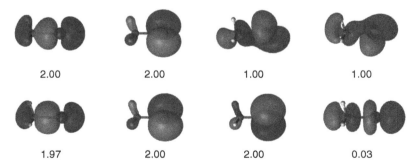

Figure 13.40   The state-specific singlet SA-CASSCF//ANO-VTZP natural orbitals and associated occupation number of CH_3Br, at the CASPT2//ANO-VDZP equilibrium structure, for the first doubly degenerate excited states (top panel) and the ground state (lower panel).

Figure 13.41 presents the natural orbitals of the same two states at a CH_3-Br distance of 4.00 Å. While the occupation numbers of the excited states have not changed significantly, we note that the σ/σ^* occupation has changed from 1.97/0.03 to 1.20/0.80 – as we dissociate the molecular system the σ and σ^* orbitals become degenerate and at infinite separation we would expect equal occupation numbers. On the other hand, the shape of the orbitals changes considerably, in the sense that the bonding and antibonding σ are quite different at the equilibrium structure, but at dissociation, they differ just in the relative phase of the two atomic orbitals. We also note that at large CH_3-Br separation there is no significant difference between the three 4p orbitals of bromine, apart from direction.

Finally, in Figure 13.42, we present the potential energy curves at the spin-free and spin-orbit level of theory. Here we note that, while the spin-free curves completely lacks the $P_{1/2}$ and $P_{3/2}$ splitting in the asymptote of the dissociation channel the spin-orbit results nicely depicts the expected splitting and at 4.00 Å predicts it to be 9.69 kcal/mol (Experiment: 10.54 kcal/mol).

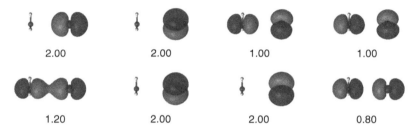

Figure 13.41 The state-specific singlet SA-CASSCF//ANO-VTZP natural orbitals and associated occupation number of CH_3Br, at a CH_3-Br bond distance of 4.00 Å, for the first doubly degenerate excited states (top panel) and the ground state (lower panel).

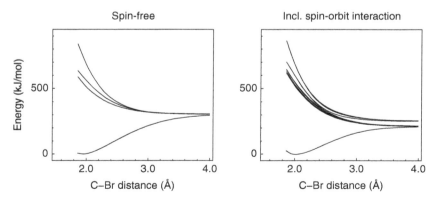

Figure 13.42 The lowest CASPT2 potential energy curves of CH_3Br along the reaction coordinate corresponding to constrained optimizations with fixed CH_3-Br bond distances.

We note that the quantitative accuracy of this model can be improved by increasing the active space, especially on the bromine. Inclusion of the electronic shells before the valence shells and the first set of virtual shells will be instrumental in achieving quantitative accuracy.

Summary We have, in this section, demonstrated the process of designing spin-orbits calculations using the MS-CASPT2/SA-CASSCF/SO-RASSI paradigm for the study of chemical reactions. In most aspects, the same philosophy is applied as for standard spin-free MS-CASPT2/CASSCF studies. However, in the SO cases, additional states of different spin-multiplicities have to be considered. Normally, these states can be described with the same active space. Special considerations might have to be considered in fragmentation channels producing atomic species – spatial degeneracies. SO-calculations on higher excited states are complicated by the presence of Rydberg states.

13.2.6 Lanthanide Chemistry

Introduction The lanthanide series consists of the 15 elements from $Z = 57$ La, which has the valence electron configuration $(4f)^0(6s)^2(5d)^1$, up to and including element $Z = 71$ Lu, with configuration $(4f)^{14}(6s)^2(5d)^1$. They form trivalent ions, collectively called Ln^{III}, and the ions have quite naturally the general valence configuration $(4f)^n$ with $n = 0, \ldots, 14$. Generally speaking, the electronic structure of lanthanide compounds can be difficult to compute, but this depends on which ligands and which states that should be treated. In particular, the later lanthanides tend to have a large number of low-lying states. This is the main fact that makes a multiconfigurational treatment mandatory. The small energy differences may require a large basis set. In addition, the large spin-orbit coupling effects require a relativistic treatment, and it is not advised to mix relativistic basis sets for the metal with nonrelativistic basis functions on ligands.

A series of calculations that were not very tasking is reported in an article by Roos and Pyykkö [37], on some trends in double-bonding properties. The calculations were aimed at filling in some gaps in standard atomic radii, but by serendipity also revealed a rather extreme case of so-called *agostic* interaction, between a metal center and an atom belonging to the ligands: This is the case of the ion $CeCH_2^+$, which breaks the expected symmetry and forms one short, and one long, nonbonded distance between the cerium and the hydrogens.

The agostic interaction breaks the C_{2v} symmetry while the C_s symmetry is retained. The HCH subunit does not change much, but is strongly tilted sideways in the symmetry plane, while the CeC distance is virtually unchanged.

The ions $LnCH_2^+$ have a double bond to methylene, except for Ln=Eu and Ln=Yb, where the extra stability of the f^7 and f^{14} shells force the single-bonded structure to be the ground state. For the purpose of the above-mentioned article, a double-bonded structure analogous to that of the other ions was found as an excited state.

The lanthanides, in particular the later ones, tend to have a large number of low-lying states. This is the main fact that makes a multiconfigurational treatment mandatory. In addition, the large spin-orbit coupling effects require a relativistic treatment.

The Active Space   The active space consisted of four electrons in four orbitals, two antibonding and two bonding, to describe the double bond to methylene, and in addition the seven f orbitals, with various number of electrons as appropriate for the specific ions. In some cases, additional active orbitals were used; see the article [37] for details.

One outcome of the calculations was a series of ion radii [38], and the values of the f-electron promotion energy (negative in the cases of Eu and Yb) and bond-breaking energies. An unforeseen discovery was the exceptional agostic interaction displayed by the $CeCH_2^+$.

For the first few ions in the series, the active space chosen is overkill: much fewer orbitals could have been used, but the decision was of course made with an eye at using as far as possible a consistent set of calculations across the entire series. For the case of the $CeCH_2^+$ ion, at the equilibrium structure, the first five natural active orbitals are shown in Figure 13.43.

As seen in the figure, their occupation numbers are 1.95, 1.92, 1.00, 0.06, and 0.04. The ones not displayed in the figure are of less interest.

In the figure is clearly seen the strong interaction between a d-orbital of Ce and one of the H atoms, which as a result is dragged closer to the metal. The direct interaction of this orbital is with the carbon, forming a σ bond. The next orbital is the π-bonding orbital, and these two form the double bond to the methylene. The effective bond order is EBO $= 1.88$, a figure that remains virtually constant across the whole series of compounds, one of the indications that the various results are comparable. For this

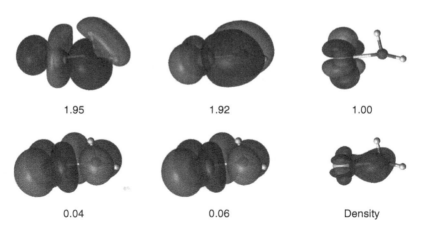

Figure 13.43   The active orbitals of a CASSCF calculation for the $CeCH_2^+$ molecular ion, using 5 active electrons in 11 orbitals. Top row: strongly occupied orbitals. Bottom left: most important correlating orbitals. Bottom right: electron density isosurface at level 0.28 au.

compound, the next orbital of importance is the singly occupied, nonbonding $\phi(f)$ orbital. The figure was obtained from a recomputation of the original one, which was made only for the ground state. If optimized for an average of several states, several more of the active orbitals could have been employed in the calculation, and a number of quite low-lying states may be expected.

Summary The "natural" choice of active space for the lanthanide compounds is sufficient, and even ample, as long as there is not too many extra orbitals needed for the ligands.

13.2.7 Actinide Chemistry

The Uranium Dimer

Introduction Based on a study using CASSCF, CASPT2, and SO-RASSI, Gagliardi and Roos [39] reported in a letter to Nature that the Uranium dimer, U_2, has a quintuple bond with a unique electronic structure.
  To quote the abstract,

> ... the molecule contains three electron-pair bonds and four one-electron bonds (that is, 10 bonding electrons, corresponding to a quintuple bond), and two ferromagnetically coupled electrons localized on one U atom each – so all known covalent bonding types are contributing.

This was followed up by a more extensive study [40] on Ac_2, Th_2, Pa_2, and U_2, showing systematic and individual features of the bonding in these dimers. The U_2 study published in Nature was confirmed but refined. These calculations suggested that the ground state was a septet ($S = 3$) with very large electronic angular momentum along the axis. However, there were obviously a multitude of low states, and the study was extended by including a larger set of states, and also the quintets.
  Using the SO-RASSI method, the spin-orbit coupling is treated as an added term in the Hamiltonian, using precomputed spin-free states as wave function basis. The spin-orbit term will couple wave functions with different angular momenta, and neither Λ nor S are good quantum number any more. The new set of calculations included 38 septet states within 0.36 eV (2930 cm^{-1}) and 32 quintet states within 0.78 eV (6290 cm^{-1}) of the ground state energy, as computed by the CASSCF/CASPT2 method. The calculations were performed using the C_{2h} point group symmetry, since this results in exact degeneracy of states with $\Lambda > 0$. Counting the spin multiplicity, this will result in $7 \times 38 + 5 \times 32 = 426$ individual spin-orbit components. However, the inversion symmetry remains a good symmetry, and the actual number of spin-orbit states in the secular equation was 386.

The Active Space The atomic 5f, 6d, and 7s orbitals give 26 molecular orbitals, where the three $7s\sigma_g$ and $6d\pi_u$ orbitals are strongly occupied at bonding distances. In this study, they can be left inactive, and their antibonding counterparts are then

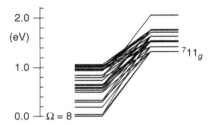

Figure 13.44   The lowest energy levels of U_2 at 2.43 Å bond length, computed with (left) and without (right) spin-orbit interaction. Energies are relative to the $\Omega = 8$ ground state.

irrelevant for the CASSCF. This leaves an active space formed by six electrons distributed among the 20 remaining orbitals.

Summary   The electronic ground state cannot be safely determined from this study: the lowest state is clearly a septet $(S = 3)$ state, but the two lowest states have contributions from both $\Lambda = 11$ and $\Lambda = 12$, and the total (orbital+spin) angular momentum along the axis is $\Omega = 8$ and $= 9$, respectively, for the two lowest states (Figure 13.44).

Their minima differ by only 0.01 eV, and it is not possible then to say with certainty which one is the lowest. Regardless of this, the bonding can be characterized as consisting of a triple bond, nominally $(s\sigma_g)^2(6d\pi_u)^4$ distributed over a large number of configuration state functions. About 96% weight of the CASSCF wave function belong to one of the configurations with occupation $(d\sigma_g)^1(d\delta_g)^1$ $(f\delta_g)^1(f\pi_u)^1(f\phi_u)^1(f\phi_g)^1$ (61%) or $(d\sigma_g)^1(d\delta_g)^1(f\delta_u)^1(f\pi_g)^1(f\phi_u)^1(f\phi_g)^1$ (35%).

The effective bond order is 4.2, and the diatom can be regarded as having a formally quintic bond, but the dissociation energy is only $D_e = 1.15$ eV since the π bonds are not fully formed and the single-electron bonds $(d\sigma_g)^1(d\delta_g)^1$ are weak.

UC and CUC

Introduction   In a combined experimental and theoretical work [41], Wang et al. reported calculations showing CUC being linear with two equivalent triple bonds, with EBO 2.83 and bond length $r_e = 1.840$ Å. The triple bond of UC had a similar EBO $= 2.82$, and a bond length $r_e = 1.855$ Å.

These results were entirely consistent with the spectroscopic results, taken from a solid argon matrix at 8 K temperature, with isotope substitution (^{12}C and ^{13}C).

The Active Space   The initial active spaces were of size 12 or 14 orbitals, and calculations were done with the ANO-RCC-VTZP basis sets. The systems included UC, CUC, and CUC+, of which the latter two were using the Cs symmetry, while UC was done using C_2. For the final calculations, the basis was upgraded to ANO-RCC-VQZP. The energies of the spin-free CASSCF states were corrected by CASPT2. In a final step, the wave function basis was expanded by using all spin-components of the spin-free states, and AMFI spin-orbit interaction terms were added to the resulting Hamiltonian matrix. This final calculation used the CASSCF states for computing the spin-orbit interaction terms but used the spin-free energies corrected by CASPT2.

For the orbital figures (Figures 13.45 and 13.46), the calculations were repeated using 10 electrons in 14 orbitals (UC) and 12 electrons in 12 orbitals (CUC). The latter calculation was at the ground state equilibrium structure, using C_2 symmetry.

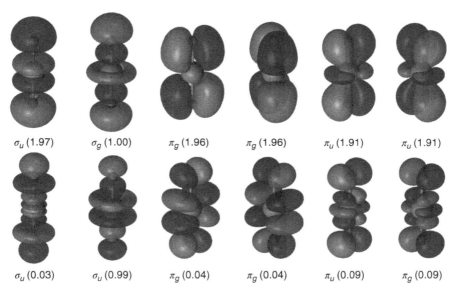

σ_u (1.97) σ_g (1.00) π_g (1.96) π_g (1.96) π_u (1.91) π_u (1.91)

σ_u (0.03) σ_u (0.99) π_g (0.04) π_g (0.04) π_u (0.09) π_g (0.09)

Figure 13.45 The CASPT2 natural orbitals of CUC (linear) for the $^3\Sigma_0^+$ ground state, with occupation numbers.

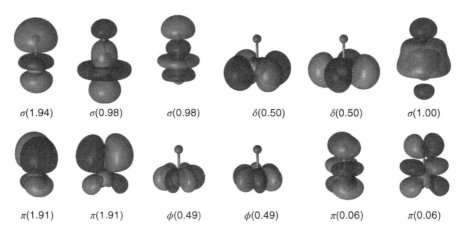

σ(1.94) σ(0.98) σ(0.98) δ(0.50) δ(0.50) σ(1.00)

π(1.91) π(1.91) ϕ(0.49) ϕ(0.49) π(0.06) π(0.06)

Figure 13.46 The CASPT2 natural orbitals of UC for the ground state ($S = 2, \Lambda = 5, \Omega = 3$), with (averaged) occupation numbers.

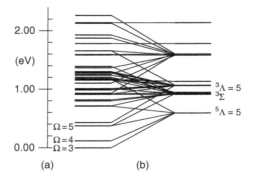

Figure 13.47 The lowest electronic energy levels of UC at $r_e = 1.855 \, \text{Å}$, computed with spin-orbit interaction included (a) and not included (b). Energies are relative to the $\Omega = 3$ ground state. There are 72 spin-orbit states in the figure.

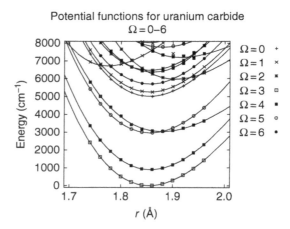

Figure 13.48 The potential functions of the lowest few spin-orbit states of UC.

The orbitals shown are the CASSCF natural orbitals for CUC, since the state is not degenerate, but for the UC it is the average natural orbitals from CASPT2.

The electronic levels are very dense, as shown in Figures 13.47 and 13.48.

Summary For these small molecules, it is relatively easy to find a good active space. On the other hand, the multitude of states that must be included makes it difficult to make calculations accurate enough to determine the exact ground state with certainty.

13.2.8 RASSCF/RASPT2 Applications

In many systems, adiabatic energy functions of excited states, and sometimes lowest states, have many weak avoided crossings or even conical intersections and cannot be reasonably described without a multiconfigurational method. Also, the multiplet structure resulting from open atomic shells can give rise to dense sets of electronic states in, for example, transition metal complexes.

As described earlier, a standard solution is to use a CASSCF/CASPT2 method, where the CASSCF is capable of describing the nondynamic correlation sufficiently well such that the subsequent CASPT2 can treat the dynamic correlation by perturbation theory.

The "Complete Active Space" description means that in principle all configuration state functions are included, that have a certain number of active electrons distributed among the active orbitals. This is a very convenient way of specifying the possible electronic structures, and also has technical advantages: the CI codes can be highly efficient, and the wave function optimization, while not always free of problems, is not cursed by the ubiquitous second minima, symmetry breaking, and numerical catastrophes that commonly afflicts general MCSCF. However, these advantages come at a price in computational resources; the number of CSFs in a typical application, having 12 active electrons and 12 active orbitals, is less than a quarter of a million, and can be handled routinely. Correlating one more electron pair gives more than 12 times more CSFs.

In the attempt to use the same, or at least a similar, active space for all the states and conformations under study, the limitations in computer time and/or memory usually dictates that one ends up in a compromise with roughly this size.

A large active space will consist of a number of orbitals that are almost doubly occupied, and a number of correlating orbitals with low occupancy, and only a few orbitals will (normally) have great variability in their occupation. Moreover, while different orbitals do not have independently varying occupation, it is nevertheless the case that (normally) the many CSFs where many electrons occupy correlating orbitals play a negligible role in describing the electronic structure.

The RASSCF wave function is thus constructed by restricting the occupations of the active orbitals. The active orbitals space is divided into three subspaces; the RAS1 space is spanned by strongly occupied orbitals, and a limit is set on the number of electrons that are allowed to leave this space. This limit is called the "maximum number of holes" that are allowed in the RAS1 space. Similarly, there is a RAS3 space of correlating orbitals, and a maximum number of electrons is allowed in this space. The intermediate RAS2 space has no such restrictions.

As an example, allowing up to two RAS1-holes and up to two RAS3-electrons would be a fairly common choice. If the number of orbitals in RAS3 were large, this would in fact be a multiconfigurational CI wave function, restricted to single and double replacements relative to a reference space of CAS type, but with optimized orbitals. It is often called precisely that, that is, a "singles and doubles type" RASSCF, SD-RASSCF, or similar. Similarly, one can define up to (say) "SDTQ56" RASSCF calculations with up to six RAS1-holes and six RAS3-electrons, but of course other combinations are possible and not uncommon.

RASSCF programs have been available for quite some time. The method works, and may be a good alternative, sometimes the only alternative, to a CASSCF. At the same time, it is a step toward general MCSCF, and has the expected problems: loss of size consistency, slower convergence, and generally less efficient codes, as compared to CASSCF. The slower convergence can be attributed to the fact that the wave function is almost, but not quite, invariant to orbital rotations between RAS1 and strongly occupied RAS2 orbitals, or between RAS3 and weakly occupied RAS2 orbitals, when these are combined with appropriate variation of the CI coefficients. This renders such orbital rotations very difficult to optimize. The performance of an optimization algorithm depends on the properties of the second derivative of the energy w.r.t the orbital and CI parameters, the so-called Hessian matrix. Corresponding to the almost-invariance property mentioned above, the Hessian has some eigenvalues that are small and difficult to model accurately. This problem becomes more pronounced if increased excitation levels are allowed, and it is commonly accepted to keep the orbitals optimized by an SD-RASSCF and performing an SDTQ-RASCI without further orbital optimization.

A good rule is to use as RAS3 orbitals only those that have small occupation in every state of importance to the calculation, and similarly RAS1 orbitals should never be allowed to have average occupation deviating much from 2.0. When doing RASSCF calculations for the first time on some new system, it happens that an RAS3 orbital is occupied in some excited state, but the relative energy of that state, compared to, for example, the ground state, would probably not be very accurate. For special purposes, such calculations have been profitably used to obtain Rydberg orbitals, in order to describe the mixing of valence and Rydberg states, or for obtaining a second d-shell for transition metal complexes [21]. RAS subspaces can be also be used, for example, in calculations on core-hole excited states.

RASPT2 Adding perturbation theory on top of a true RASSCF wave functions is not only easy but also requires additional calculations that are not needed for CASPT2. In particular, a straightforward implementation in the same way as CASPT2 would require computing a four-body density matrix (with active indices only), and another matrix of the same size. These tensorial quantities, with eight active indices, grow very quickly with the number of active orbitals, which conflicts with the idea of using RASPT2 to allow larger active spaces. On the other hand, these quantities are needed only for perturbation contributions to the energy due to the interaction with CSFs that would be present in the CASSCF but are excluded by the RAS restrictions. By analogy to conventional truncated CI, such interactions could be regarded as primarily due to unlinked products of the lower excitations and would not be very important energetically in a perturbation approach. Thus, it was decided to postpone the inclusion of excitations that are completely within the active space but outside of the restricted space. Once more experience with this approximation has been accumulated, it might be seen whether it would pay off to do a full implementation.

An early RASPT2 program was included in the program package MOLPRO [42], but it appears not to have been much used. Moreover, the variant of CASPT2 on which that program is built is different from the one used by MOLCAS. The description here is thus limited to the MOLCAS implementation. Our first application was to mono- and binuclear copper complexes [43], and used up to 24 active electrons in 28 orbitals, large enough to give useful results while clearly out of reach for any CASSCF calculation.

Huber et al. [44] applied the RASSCF method to five intermediates in the reaction of O_2 with a Cu(I)-complex, for which they were able to do CASPT2 calculations with 12 electrons in 12 active orbitals (12,12). It was shown that the CSF space could be restricted by two orders of magnitude, by using RAS restrictions with a minimal RAS2 space to account for nondynamic correlation, and the rest of the 12 active orbitals used for a RAS1 and a RAS3 space, such that the reference RASSCF wave function became equivalent to a multireference singles and doubles CI, with essentially the same accuracy as the larger CASPT2 calculation.

A fairly large benchmark study has been published by Sauri et al. [21]. They computed excited states of a number of organic and inorganic molecules, including, for example, free base porphin, nucleobases, and the copper tetrachloride dianion, using several different strategies for the restriction of the active space, and compared to CASPT2, CCSD, and experimental data.

Ruiperez et al. [45] applied RASPT2 to Cr_2 with good results. In this study, they also found that the IPEA shift should be used with a significantly larger parameter than commonly used (0.45 instead of 0.25). This has also been noted in a CASPT2 study of spin states of iron and cobalt complexes [46]. However, later studies [47] suggest that the demand for large IPEA shifts was exaggerated by a too small active space.

Excited states of chromium and iron complexes CrF_6, ferrocene, $Cr(CO)_6$, ferrous porphin, cobalt corrole, and FeO/FeO^- have been studied by Vancoillie et al. [48], using RASSCF/RASPT2. This study showed, for example, how the high magnetic moment of the $^3A_{2g}$ ground state of Fe(II)porphin was caused by spin-orbit interaction with nearby states. These are very demanding systems, in particular the cobalt corrole, which they reported needed 25–33 active orbitals. Comparison between RASPT2 and range-separated hybrid DFT have been reported by Escudero and Gonzalez [49] for a Ru(II)bipyridyl complex, and between RASPT2/SO-RASSI and TDDFT for UO_2 by Wei et al. [50].

Instead of specifically using the RAS restriction format, with its three active subspaces, it has been suggested that the more general GAS (Generalized Active Space) could be used as a so-called Split-GAS scheme [51]. The GAS concept implies that a number (not prescribed) of subspaces are used, each with a maximum and a minimum occupation, to allow a flexible and orbital-based description of suitable MCSCF calculations for, for example, transition metals. The Split-GAS scheme is a method that allows in principle larger active spaces with less computational expense. Other suggestions are to use so-called DMRG wave functions: this stands for Density Matrix Renormalization Group [52, 53], a concept originally borrowed from the study of spin lattices. It seems clear that further increase of the feasible size of active spaces is necessary, and that this demands a flexible parameterization of the wave function by some tensor factor decomposition.

13.3   REFERENCES

[1] Pipek J, Mezey PG. A fast intrinsic localization procedure applicable for ab initio and semiempirical linear combination of atomic orbital wave functions. J Chem Phys 1989; 90:4916–4926.

[2] Pulay P. Localizability of dynamic electron correlation. Chem Phys Lett 1983; 100:151–154.

[3] Saebo S, Pulay P. Local treatment of electron correlation. Annu Rev Phys Chem 1993; 44:213–236.

[4] Aquilante F, Pedersen TB, Sánchez de Merás A, Koch H. Fast noniterative orbital localization for large molecules. J Chem Phys 2006; 125:Art. 174101.

[5] Edmiston C, Ruedenberg K. Localized atomic and molecular orbitals. Rev Mod Phys 1963; 35:457–464.

[6] Boys SF. Construction of some molecular orbitals to be approximately invariant for changes from one molecule to another. Rev Mod Phys 1960; 32:296–299.

[7] Jones RR, Bergman RG. p-Benzyne – Generation as an intermediate in a thermal isomerization reaction and trapping evidence for 1,4-benzenediyl structure. J Am Chem Soc 1972; 94:660–661.

[8] Bergman RG. Reactive 1,4-dehydroaromatics. Acc Chem Res 1973; 6:25–32.

[9] Lockhart TP, Comita PB, Bergman RG. Kinetic evidence for the formation of discrete 1,4-dehydrobenzene intermediates - trapping by intermolecular and intramolecular hydrogen-atom transfer and observation of high-temperature CIDNP. J Am Chem Soc 1981; 103:4082–4090.

[10] Lockhart TP, Bergman RG. Evidence for the reactive spin state of 1,4-dehydrobenzenes. J Am Chem Soc 1981; 103:4091–4096.

[11] Lindh R, Persson BJ. Ab initio study of the Bergman reaction: the autoaromatization of Hex-3-ene-1,5-diyne. J Am Chem Soc 1994; 116:4963–4969.

[12] Lindh R, Lee TJ, Bernhardsson A, Persson BJ, Karlström G. Extended ab initio and theoretical thermodynamics studies of the Bergman reaction and the energy splitting of the singlet o-, m-, and p-Benzynes. J Am Chem Soc 1995; 117:7186–7194.

[13] Finley J, Malmqvist PÅ, Roos BO, Serrano-Andrés L. The multi-state CASPT2 method. Chem Phys Lett 1998; 288:299–306.

[14] Rubio M, Serrano-Andrés L, Merchán M. Excited states of the water molecule: analysis of the valence and Rydberg character. J Chem Phys 2008; 128:Art. 104305.

[15] Bigelow RW, Freund HJ, Dick B. The T_1 state of $para$-nitroaneline and related molecules - a CNDO/S study. Theor Chim Acta 1983; 63:177–194.

[16] Wortmann R, Kramer P, Glania C, Lebus S, Detzer N. Deviations from Kleinman symmetry of the 2nd-order polarizability tensor in molecules with low-lying perpendicular electronic bands. Chem Phys 1993; 173:99–108.

[17] Farztdinov VM, Schanz R, Kovalenko SA, Ernsting NP. Relaxation of optically excited p-nitroaniline: semiempirical quantum-chemical calculations compared to femtosecond experimental results. J Phys Chem A 2000; 104:11486–11496.

[18] Millefiori S, Favini G, Millefiori A, Grasso D. Electronic-spectra and structure of nitroanilines. Spectrochim Acta, Part A 1977; 33:21–27.

[19] Roos BO, Fülscher MP, Malmqvist PÅ, Merchán M, Serrano-Andrés L. Theoretical studies of electronic spectra of organic molecules. In: Langhoff SR, editor. Quantum

Mechanical Electronic Structure Calculations with Chemical Accuracy. Dordrecht, The Netherlands: Kluwer Academic Publishers; 1995. p. 357–438.

[20] Kaufmann K, Baumeister W, Jungen M. Universal Gaussian basis sets for an optimum representation of Rydberg and continuum wave functions. J Phys B: At Mol Opt Phys 1989; 22:2223–2240.

[21] Sauri V, Serrano-Andrés L, Shahi ARM, Gagliardi L, Vancoillie S, Pierloot K. Multiconfigurational second-order perturbation theory restricted active space (RASPT2) method for electronic excited states: a benchmark study. J Chem Theory Comput 2011; 7:153–168.

[22] De Vico L, Liu YJ, Krogh JW, Lindh R. Chemiluminescence of 1,2-dioxetane. Reaction mechanism uncovered. J Phys Chem A 2007; 111:8013–8019.

[23] Moriarty N, Lindh R, Karlström G. Tetramethylene: a CASPT2 study. Chem Phys Lett 1998; 289:442–450.

[24] Navizet I, Liu YJ, Ferré N, Roca-Sanjuán D, Lindh R. The chemistry of bioluminescence: an analysis of chemical functionalities. Comput Phys Commun 2012; 12:3064–3076.

[25] Serrano-Andrés L, Merchán M. Are the five natural DNA/RNA base monomers a good choice from natural selection? A photochemical perspective. J Photochem Photobiol C 2009; 10:21–32.

[26] Matsika S, Spanner M, Kotur M, Weinacht TC. Ultrafast relaxation dynamics of uracil probed via strong field dissociative ionization. J Phys Chem A 2013; 117:12796–12801.

[27] Wang Y, Haze O, Dinnocenzo JP, Farid S, Farid R, Gould I. Bonded exciplexes. A new concept in photochemical reactions. J Org Chem 2007; 72:6970–6981.

[28] Veryazov V, Malmqvist PÅ, Roos BO. How to select active space for multiconfigurational quantum chemistry? Int J Quantum Chem 2011; 111:3329–3338.

[29] Froese-Fischer C. Oscillator strengths for ^{2}S–^{2}P transitions in the copper sequence. J Phys B: At Mol Phys 1977; 10:1241–1251.

[30] Andersson K, Roos BO. Excitation energies in the nickel atom studied with the complete active space SCF method and second-order perturbation theory. Chem Phys Lett 1992; 191:507–514.

[31] Roos BO, Veryazov V, Conradie J, Taylor PR, Ghosh A. Not innocent: verdict from ab initio multiconfigurational second-order perturbation theory on the electronic structure of chloroiron corrole. J Phys Chem B 2008; 112:14099–14102.

[32] Roos BO, Borin AC, Gagliardi L. Reaching the maximum multiplicity of the covalent chemical bond. Angew Chem Int Ed 2007; 46:1469–1472.

[33] La Macchia G, Aquilante F, Veryazov V, Roos BO, Gagliardi L. Bond length and bond order in one of the shortest Cr-Cr Bonds. Inorg Chem 2008; 47:11455–11457.

[34] Shafirovich V, Lymar SV. Spin-forbidden deprotonation of aqueous nitroxyl (HNO). J Am Chem Soc 2003; 125:6547–6552.

[35] Harvey JN. Spin-forbidden reactions: computational insight into mechanisms and kinetics. WIREs Comput Mol Sci 2014; 4:1–14.

[36] Pyykkö P. Relativistic effects in structural chemistry. Chem Rev 1988; 88:563–594.

[37] Roos BO, Pyykkö P. Bonding trends in molecular compounds of lanthanides: the double-bonded carbene cations $LnCH_2^+$ (Ln=Sc,Y,La–Lu). Chem Eur J 2010; 16:270–275.

[38] Pyykkö P, Atsumi M. Molecular double-bond covalent radii for elements Li-E112. Chem Eur J 2009; 15:12770–12779.

[39] Gagliardi L, Roos BO. Quantum chemical calculations show that the uranium molecule U_2 has a quintuple bond. Nature 2005; 433:484.

[40] Roos BO, Malmqvist PÅ, Gagliardi L. Exploring the actinide-actinide bond: theoretical studies of the chemical bond in Ac_2, Th_2, Pa_2, and U_2. J Am Chem Soc 2006; 128:17000–17006.

[41] Wang X, Andrews L, Malmqvist PÅ, Roos BO, Gonçalves AP, Pereira CCL, Marçalo J, Godart C, Villeroy B. Infrared spectra and quantum chemical calculations of the uranium carbide molecules UC and CUC with triple bonds. J Am Chem Soc 2010; 132:8484–8488.

[42] Werner HJ, Knowles PJ, Knizia G, Manby FR, Schütz M, Celani P, Korona T, Lindh R, Mitrushenkov A, Rauhut G, Shamasundar KR, Adler TB, Amos RD, Bernhardsson A, Berning A, Cooper DL, Deegan MJO, Dobbyn AJ, Eckert F, Goll E, Hampel C, Hesselmann A, Hetzer G, Hrenar T, Jansen G, Köppl C, Liu Y, Lloyd AW, Mata RA, May AJ, McNicholas SJ, Meyer W, Mura ME, Nicklass A, O'Neill DP, Palmieri P, Peng D, Pflüger K, Pitzer R, Reiher M, Shiozaki T, Stoll H, Stone AJ, Tarroni R, Thorsteinsson T, Wang M. MOLPRO, Version 2012.1, a Package of ab Initio Programs. Germany: TTI GmbH, Stuttgart; 2015. See http://www.molpro.net.

[43] Malmqvist PÅ, Pierloot K, Shahi ARM, Cramer CJ, Gagliardi L. The restricted active space followed by second-order perturbation theory method: theory and application to the study of CuO_2 and Cu_2O_2. J Chem Phys 2008; 128:Art. 204109.

[44] Huber SM, Shahi ARM, Aquilante F, Cramer CJ, Gagliardi L. What active space adequately describes oxygen activation by a late transition metal? CASPT2 and RASPT2 applied to intermediates from the reaction of O_2 with a Cu(I)-r-Ketocarboxylate. J Chem Theory Comput 2009; 5:2967–2976.

[45] Ruipérez F, Aquilante F, Ugalde JM, Infante I. Complete vs restricted active space perturbation theory calculation of the Cr_2 potential energy surface. J Chem Theory Comput 2011; 7:1640–1646.

[46] Daku LML, Aquilante F, Robinson TW, Hauser A. Accurate spin-state energetics of transition metal complexes. 1. CCSD(T), CASPT2, and DFT study of $[M(NCH)_6]^{2+}$ (M = Fe, Co). J Chem Theory Comput 2012; 8:4216–4231.

[47] Vancoillie S, Malmqvist PÅ, Veryazov V. Potential energy surface of the chromium dimer re-re-revisited with multiconfigurational perturbation theory. J Chem Theory Comput 2016; 12:1647–1655.

[48] Vancoillie S, Zhao H, Tran VT, Hendrickx MFA, Pierloot K. Multiconfigurational second-order perturbation theory restricted active space (RASPT2) studies on mononuclear first-row transition-metal systems. J Chem Theory Comput 2011; 7:3961–3977.

[49] Escudero D, González L. RASPT2/RASSCF vs range-separated/hybrid DFT methods: assessing the excited states of a Ru (II) bipyridyl complex. J Chem Theory Comput 2011; 8(1): 203–213.

[50] Wei F, Wu GS, Schwarz WHE, Li J. Excited states and absorption spectra of UF_6: a RASPT2 theoretical study with spin-orbit coupling. J Chem Theory Comput 2011; 7:3223–3231.

[51] Manni GL, Ma D, Aquilante F, Olsen J, Gagliardi L. SplitGAS method for strong correlation and the challenging case of Cr_2. J Chem Theory Comput 2013; 9:3375–3384.

[52] Schollwöck U. The density-matrix renormalization group. Rev Mod Phys 2005; 77:259–315.

[53] Hachmann J, Dorando JJ, Avilés M, Chan GKL. The radical character of the acenes: a density matrix renormalization group study. J Chem Phys 2007; 127:Art. 134309.

SUMMARY AND CONCLUSION

This book's intention is to demystify the multiconfigurational wave function model and how it is used in computational chemistry. The book takes the reader through the theory of the method and with a large array of applications, in different fields of applied chemistry, demonstrates proper use. We, the authors, hope that the book will assist in popularizing the method and increase its use in everyday computational chemistry. In particular, we note that recent developments, as new forms of parameterization and combination of the method with Monte Carlo techniques, show very promising performance in removing some of the limitations of the method, as for example, limitations with respect to how large molecular systems the method can be applied to. These developments lead us to believe that the method will again become a popular and standard tool in quantum chemical studies of molecular phenomena.

So, dear reader, if you have reached this final page of the book, you might think that the long journey is over ... Well, Yes and No. Finishing a cookbook (even a good one) does not immediately make you a good master-chef. You learned some recipes, but there are so many varieties in ingredients you might find, that you have to develop your own skills and your own recipes. And the only way to do that is to try!

Finally, any constructive feedback is welcomed.

INDEX

Multiconfigurational Quantum Chemistry, First Edition.
Björn O. Roos, Roland Lindh, Per Åke Malmqvist, Valera Veryazov, and Per-Olof Widmark.
© 2016 John Wiley & Sons, Inc. Published 2016 by John Wiley & Sons, Inc.